NORTH CAROLINA'S HURRICANE HISTORY

JAY BARNES

The University of North Carolina Press

Chapel Hill & London

NORTH CAROLINA'S
HURRICANE
HISTORY

THIRD EDITION

© 1995, 1998, 2001
The University of
North Carolina Press
All rights reserved
Manufactured in the United
States of America

The paper in this book meets
the guidelines for permanence
and durability of the Commit-
tee on Production Guidelines
for Book Longevity of the
Council on Library Resources.

Library of Congress
Cataloging-in-Publication Data
Barnes, Jay.
North Carolina's hurricane
history / Jay Barnes. —
Third ed.
    p.   cm.
Includes bibliographical
references and index.
ISBN 0-8078-2640-5
(cloth: alk. paper) —
ISBN 0-8078-4969-3
(pbk.: alk. paper)
    1. Hurricanes — North
Carolina — History.   I. Title:
Hurricane history.   II. Title.
QC945.B37 2001   2001027418
363.34'922'09576 — dc21   CIP

05 04 03 02 01   5 4 3 2 1

Design by April Leidig-Higgins
Graphics by Jackie Johnson

Title page: The awesome fury of hurricane Hazel is evident in this photo by Hugh Morton, taken near the peak of the storm at Carolina Beach. Morton and several other reporters witnessed the destruction, including *Charlotte News* staffer Julian Scheer, seen here struggling against the rising tide. This photo won Morton first place in spot news in the 1955 Southern Photographer of the Year competition. (From *Making a Difference in North Carolina*, by Ed Rankin and Hugh Morton; photo courtesy of Hugh Morton)

Page v: The people of North Carolina have always managed to show great spirit in coping with the turmoil caused by major hurricanes. (Photo courtesy of Roy Hardee)

Page vii: Wrightsville Beach Yacht Club after hurricane Hazel, October 1954. (Photo courtesy of the N.C. Division of Archives and History)

Page ix: A young girl wades through the streets of Carolina Beach after the passing of hurricane Hazel in October 1954. (Photo courtesy of Hugh Morton)

CONTENTS

Introduction 1

1 BIRTH OF A HURRICANE 5

2 HURRICANE EFFECTS 9

Winds 10

Storm Surge 14

Rainfall 18

Tornadoes 19

Storm Intensity 20

Other Factors 23

3 WATCHING THE STORMS 27

What's in a Name? 31

4 EARLY NORTH CAROLINA
HURRICANES, 1524–1861 33

5 TAR HEEL TRAGEDIES,
1875–1900 39

September 17, 1876 40

August 18, 1879 42

September 9, 1881 46

1882 46

September 11, 1883 47

August 25, 1885 47

August 27–29, 1893 48

October 13, 1893 48

August 16–18, 1899 49

October 30–31, 1899 57

6 HURRICANES OF
THE NEW CENTURY,
1900–1950 63

November 13, 1904 64

September 17, 1906 64

September 3, 1913 64

July 14–16, 1916 66

September 18–19, 1928 66

October 1–2, 1929 66

August 22–23, 1933 67

September 15–16, 1933 68

September 18, 1936 73

August 1, 1944 75

September 14, 1944 76

7 HURRICANE ALLEY,
1950–1960 79

Barbara (August 13, 1953) 80

Carol (August 30, 1954) 80

Edna (September 10, 1954) 81

Hazel (October 15, 1954) 82

Connie (August 12, 1955) 108

Diane (August 17, 1955) 110

Ione (September 19, 1955) 114

Helene (September 27, 1958) 119

Donna (September 11, 1960) 120

8 THE MODERN ERA,
1960–1999 133

Ginger (September 30–
October 1, 1971) 134

Agnes (June 20–21, 1972) 136

David (September 5, 1979) 136

Diana (September 9–14, 1984) 137

Gloria (September 26–27, 1985) 143

Charley (August 17–18, 1986) 147

Hugo (September 21–22, 1989) 149

Emily (August 31, 1993) 157

Bertha (July 12, 1996) 162

Fran (September 5–6, 1996) 172

Bonnie (August 26–28, 1998) 204

Dennis (August 30–
September 5, 1999) 214

Floyd (September 16, 1999) 220

9 NOR'EASTERS 261

10 CREATURES IN THE STORM 269

11 THE NEXT GREAT STORM 277

12 HURRICANE SURVIVAL 291

Appendix 299

Acknowledgments 307

Index 311

NORTH CAROLINA'S HURRICANE HISTORY

Hurricanes. They brew themselves out of the heat of the tropics, spinning blindly across the open sea. Often they evolve into massive storms with violent winds and torrential rains. They may live for days or for weeks, and most die off harmlessly as they wander over cooler waters. Some cause widespread alarm by tracking unpredictably close to land. We plot their growth, give them names, and track their movements across the sea. But occasionally, these storms become deadly intruders as they strike our coastlines with random fury. Hurricanes, and their counterparts around the globe, are the greatest storms on earth, killing more people worldwide than all other storms combined.

Soon after the initial release of *North Carolina's Hurricane History* in June 1995, the tropical Atlantic began to boil with activity. After several years of relative quiet, forecasters at the National Hurricane Center found themselves working around the clock, tracking a steady stream of hurricanes in alphabetical sequence. As the season progressed, portions of the Caribbean felt the wrath of hurricanes Iris, Luis, and Marilyn, while Florida was pounded by Allison, Erin, and Opal. In North Carolina, coastal residents and vacationers were driven into shelters by the frightening approach of hurricane Felix, only to see this once-powerful storm veer away and spare the coast. When it was all over, 1995 had been a near-record year for tropical weather in the Atlantic, with nineteen named storms—eleven of which became hurricanes. Within the course of one season, it seemed, hurricanes had suddenly come back in style.

It was the following year, however, that brought the Tar Heel state its due. Hurricanes Bertha and Fran pummeled the same section of coast in July and September 1996, respectively, leaving frazzled nerves and enormous destruction in their wakes. Bertha's midsummer arrival was unwelcome enough, but Fran's voracious march across the state caused record losses from the beaches of New Hanover, Pender, and Onslow Counties to the Research Triangle area and beyond. Houses sank, trees crashed, and rivers spilled over their banks as the state endured its first direct hit by a major hurricane in thirty-six years. For many thousands of its victims, Fran was a first-time weather experience that established it as the new standard for North Carolina's epic storm.

But records were made to be broken. Fran was followed by still more hurricanes in the late 1990s, including the disastrous storm that would become the newest benchmark—hurricane Floyd. Approaching at the end of the millennium with horrific, category-four vigor, Floyd frightened millions of coastal residents from Key West to Norfolk. Though it weakened considerably before

Until Fran arrived in 1996, North Carolina's benchmark storm was hurricane Hazel, which lashed the state's southern beaches in October 1954. Even though Floyd and Fran caused greater destruction, Hazel was the most powerful of the three; it was the only category-four storm to strike the state during the twentieth century. The damages at Carolina Beach were extensive. (Photo courtesy of the Cape Fear Museum)

sweeping over eastern North Carolina, the storm dumped more than a foot of rain on areas already soaked by tropical storm Dennis just two weeks earlier. The resultant flooding was unprecedented. As the waters rose, thousands fled their homes, hundreds were rescued from rooftops, and more than fifty died. The destruction in the eastern third of the state was beyond what most could have imagined. Thus, in the final months of the twentieth century, Floyd claimed its place as North Carolina's greatest hurricane disaster by destroying more homes, costing more dollars, claiming more lives, and delivering more misery than any other storm of the last hundred years.

Even with Floyd and Fran fresh on their minds, many who like to talk hurricanes still have lingering memories of Hazel, the previous generation's benchmark storm on the Tar Heel coast. Others recall that just a few years ago hurricane Hugo rocked the Carolinas, leveling a sea of fallen timber across our western counties. But Floyd, Fran, Hugo, and Hazel are only part of the story. Since the days of the first European explorers, North Carolina has had a long and brutal hurricane history. Countless big storms have overwashed our coast and battered our state, and many North Carolinians have lost their lives in the desperate struggle against water and wind. Until now, few stories have been told of the many severe storms that have swept through the state. This collection of photographs and words, updated to include recent events, pieces together that history. Our personal experiences with Fran and Floyd taught us many valuable lessons, but a review of North Carolina's hurricane history may broaden our understanding of our persistent vulnerability to hurricanes.

Isolated barrier islands and broad, shallow sounds are the dominant features of the North Carolina coast. The barrier islands, including those that make up the Outer Banks, offer the mainland some protection against hurricanes and winter storms. These islands line the extensive coastline, stretching over three hundred miles from Corolla to Calabash. Through the years, hurricanes have reshaped the coast by moving massive amounts of water and sand, overwashing the barrier islands, and opening and closing inlets.

Today the coastal islands support thousands of homes and businesses and have become major tourist destinations. Growth in the coastal zone continues at a rapid pace, bringing more and more people to the ocean. In ever-increasing numbers, residents and vacationers strive to be near the water, where they can enjoy bright, sandy beaches and the smell of salt air. However, as became obvious after Bertha and Fran, many have built in areas that suffer from the advances of major hurricanes. To most, that threat is vaguely understood as an acceptable risk.

Hurricanes that strike North Carolina usually have their greatest impact along the coast. Storm tides and large waves inundate low-lying areas, sometimes flooding streets and homes and sinking boats. Many of those who have lost their lives to these storms have been trapped by rapidly rising water.

But the devastating effects of a hurricane are not limited to the shore. Large, powerful storms like Hazel, Hugo, and Fran create widespread destruction as they track inland and can bring chaos to metropolitan areas like Raleigh and Charlotte. Gusting winds may uproot large trees, snap utility poles, and inflict extensive property damage hundreds of miles from the ocean. Drenching storms like Floyd may lack strong winds but can bring disastrous flooding to inland areas. Torrential rains fall as these massive storms weaken over land, overfilling rivers and creeks and flooding highways, farms, and cities. Powerful storms striking the Gulf states, such as Opal in 1995 and Camille in 1969, have even been known to track northeast, causing devastating flooding and wind damage throughout the Appalachian Mountains.

Through the years, all parts of North Carolina have been battered by hurricanes at one time or another. Before the arrival of Bertha and Fran, though, North Carolinians had enjoyed several decades of relatively few hurricane strikes. Hugo actually made landfall over the South Carolina coast before carving a path across that state and up through Charlotte in 1989. Moderate storms like Diana in 1984 and Gloria in 1985 had come ashore in North Carolina, but overall there were few powerful midcoast hurricanes throughout the sixties, seventies, and eighties. Hurricane Donna struck Carteret County in September 1960 and was the only *major* hurricane to make a direct hit on the North Carolina coast during this period. Hurricane Emily came extremely close to landfall on the Outer Banks in 1993, close enough to cause significant destruction at Cape Hatteras. But in the 1950s, eastern North Carolina became known as "hurricane alley" when six hurricanes hit the state within seven years. In fact, hurricanes Hazel, Connie, Diane, and Ione all made landfall within a twelve-month period from October 1954 to September 1955.

Before the 1950s, hurricanes were rarely given names, and most can be recalled only by the dates of their occurrence; we know, for example, that during the twentieth century North Carolina was struck by memorable storms in 1944, 1933, 1916, and 1913. Records of hurricanes from long ago are somewhat less complete than those of more recent storms, and photographs of hurricanes from the nineteenth century are rare. Notable hurricanes from the 1800s occurred in 1899, 1883, 1879, 1856, 1846, 1842, and 1837. Many other significant storms hit the state in earlier centuries, although even fewer details of their effects are available.

Newspaper reports of early hurricanes were often delayed for days after the actual events because communications were disrupted and traveling was difficult in the storms' aftermaths. Firsthand accounts of storm damage from the remote Outer Banks sometimes never reached the major newspapers of the Piedmont, although those journals frequently ran lengthy descriptions of the destruction caused by hurricanes in Florida, Charleston, Norfolk, and New York. Few photographs of North Carolina hurricanes were taken before the 1940s, largely due to the scarcity of cameras in the mostly rural coastal communities. Fortunately, weather stations in various statewide locations kept accurate records of wind speeds, rainfall, and the barometric pressures by which these storms are measured.

This history of North Carolina's hurricanes was compiled from a wide variety of sources, including newspaper reports, historical publications, letters, National Weather Service records, and personal interviews. In some cases, stories of coastal residents and their hurricane adventures have been passed along like other down-east folklore and offer amazing portraits of destruction and survival. Photographs illustrating the impact hurricanes have had in North Carolina are historical treasures. The ones used in this book were collected from museums, newspapers, libraries, government agencies, businesses, and family albums.

Many individuals provided valuable assistance in building this collection of photographs and events. To some, it may be of greatest interest as a photographic journal and record of a great state with a stormy past. But I also hope that, if we improve our understanding of these uncommon weather events and their local effects, we will better know what to expect and how to prepare when the *next* major hurricane strikes the North Carolina coast.

BIRTH OF A HURRICANE
1

Hurricane Diana approaches the North Carolina coast in 1984, as seen from a weather satellite. (Photos courtesy of the National Weather Service)

(Page 5)
Hurricane Hugo churns toward the Carolina coast in September 1989. (Photo courtesy of the National Weather Service)

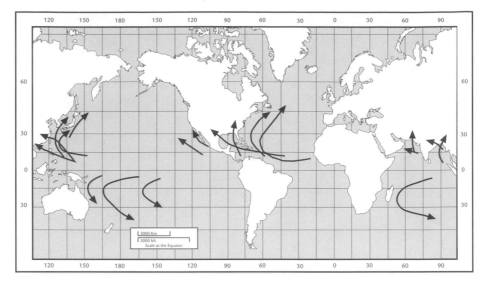

Tropical cyclone development around the world
Source: *Adapted from* Atlantic Hurricanes, *by Gordon Dunn and Banner Miller.*

There is nothing in our atmosphere that compares with their awesome fury. Arctic storms are often larger, and tornadoes may pack more violent winds, but no weather system can match the broad-scale destructive force of hurricanes. For centuries, they have left legacies of death and despair. Many tropical and temperate nations know too well the ruinous effects of these devastating storms.

In the Western Hemisphere, they are known as *hurricanes*, a term derived from the Caribbean Indian word translated as "big wind" or "storm god." *Typhoons* in the western Pacific and *cyclones* in the Indian Ocean are other names for the atmospheric phenomena we call hurricanes. All are tropical cyclones that form in the low latitudes of all tropical oceans except the South Atlantic and Southeast Pacific.

As the intense rays of the summer sun warm the ocean's surface, evaporation and conduction transfer enormous amounts of heat and moisture into the atmosphere, providing the fuel for tropical cyclones. Warm vapors rise, cool, and condense, forming billowing clouds, scattered showers, and thunderstorms. The thunderstorms grow dramatically in a passing *tropical wave*, a low-pressure trough that drifts westward through equatorial waters. The wave may develop into a *depression*, as thunderclouds build and barometric pressures drop. Under the right conditions, the depression may intensify into a tropical storm with gale-force winds and eventually become a full-grown hurricane.

Often, these storms strengthen as they begin to show signs of rotation. The earth's spin produces the *Coriolis effect*, which causes winds within the depres-

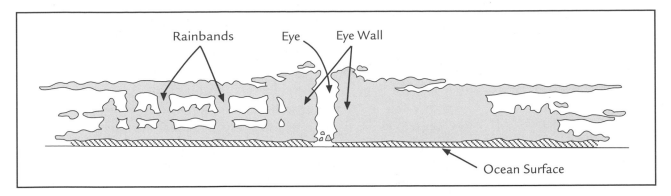

Rainbands   Eye   Eye Wall

Ocean Surface

Cross-section of a severe hurricane

Residents sift through the remains of their North Topsail Beach home in the aftermath of hurricane Bertha. (Photo by Don Bryan; courtesy of the Jacksonville Daily News)

sion to curve and bend around the central low pressure. These curving winds help intensify the storm, as warm, moist air recharges the thunderstorms. In the Northern Hemisphere, cyclones spin counterclockwise, whereas those originating below the equator spin clockwise. Once the rotation is well defined, the storm becomes organized and takes on the character of a potential hurricane. Winds near the center of the storm increase, and a relatively calm eye develops, surrounded by ominous spiral rainbands that extend outward for many miles. When sustained winds reach 74 mph or greater, the tropical storm becomes a hurricane.

A well-developed hurricane covers thousands of square miles as it drifts across the ocean's surface. Rivers of air in the atmosphere push and steer tropical storms and hurricanes. Low-level trade winds and high-altitude steering currents join to guide the storms on what are sometimes erratic courses.

Many hurricanes appear to wobble, like a child's top spun precariously across a table. As long as they remain over warm water, tropical cyclones can strengthen and grow. To intensify, they need a good supply of fuel — the heat and moisture available to the atmosphere at sea. Once they pass over land, they lose their source of energy, encounter increased friction, and begin to weaken. Often, however, hurricanes striking the coast may curve back out to sea, where they can regain their intensity and come ashore in other regions.

The heat required to fuel these storms is at its peak during the long, hot days of late summer. August and September are prime months for tropical storms in the North Atlantic, but the official hurricane season lasts from June 1 to November 30. The earliest known start for North Carolina's hurricane season was on June 3, 1825, when a major storm struck the state. One hundred years later, the latest North Carolina hurricane on record crossed Carteret County on December 2, 1925. Perhaps the most notorious of all Carolina storms, hurricane Hazel, struck in October 1954. Historically, however, most hurricanes have made landfall during the season's peak in late summer.

In an average year, we can expect more than one hundred tropical disturbances to form in the Atlantic, Gulf of Mexico, and Caribbean. Of these, only ten will reach tropical-storm intensity, and only six of these storms will become hurricanes. On average, two of these hurricanes will strike the U.S. coastline, anywhere between Texas and Maine. Some portion of North Carolina can expect to be affected by a hurricane about once every four years, based on the number of storms that have struck the state over the last century.

True hurricanes usually have tropical origins and fall within the June to November season. But other dreadful storms can strike during the winter months, featuring many of the same destructive characteristics as their tropical counterparts. These winter storms, or *northeasters*, often batter the Carolina coast with strong northeast winds that cause high tides, widespread flooding, and extensive beach erosion.

Northeasters are actually extratropical cyclones, similar in many respects to hurricanes but lacking a central warm air mass and a well-defined eye. They usually develop when low-pressure systems move out of the Gulf of Mexico and into the Atlantic, where they gain strength from the warm waters of the Gulf Stream. These winter storms are frequent visitors to the Outer Banks, sometimes stalling offshore and pounding the coast for days. The Ash Wednesday Storm of 1962 and the March Superstorm of 1993 are two classic northeasters whose destructive legacies compare with those of some of our worst hurricanes.

HURRICANE EFFECTS **2**

Violent hurricanes and north-easters can bring awesome destruction to coastal areas, as evidenced in this scene from the Ash Wednesday Storm. (Photo courtesy of Roy Hardee)

Media coverage of modern hurricanes like Fran and Floyd has provided graphic evidence of the violence unleashed by these storms. When powerful hurricanes strike populated areas the effects can be devastating. Their high winds leave homes and businesses with the bombed-out appearance of a war zone. The surging ocean and unyielding rain inundate coastal communities and inland rivers, flooding out homes, highways, and farmlands. The combined natural forces of wind and water sometimes take the lives of the unprepared and cause property losses in the billions of dollars.

But how does the hurricane machine deal out such destruction? Each storm may affect a region differently, but every hurricane is capable of striking in many ways. Forecasters look carefully at the measurable components of each storm to create an accurate picture of its potential for disaster.

WINDS

By definition, a tropical storm becomes a hurricane when its constant wind speeds are determined to be 74 mph or greater. At this minimal intensity, the hurricane may bring modest damage to trees, signs, roofs, and other structures. Wind gusts may exceed this level and create greater destruction in isolated areas near the center of the storm. More intense hurricanes pack much higher winds. Storms with sustained winds greater than 120 mph can cause more significant structural damage, as this is the threshold the North Carolina coastal building code sets as the design wind speed, the force newly constructed buildings must be able to withstand. As wind speeds increase, the forces they exert on a structure begin to multiply. The 120-mph winds of a major hurricane would exert about four times the force of a 60-mph wind.

Rarely, "super hurricanes" may develop with constant wind speeds of over 155 mph and gusts exceeding 200 mph. These tornado-speed velocities produce deadly consequences, as even substantial structures may be blown to pieces. Hurricane Camille, which struck near Biloxi, Mississippi, in August 1969, was just such a storm. It is estimated that Camille's winds were over 175 mph, although meteorologists concede that such extreme winds are difficult to verify. In some areas, gusts may have topped 200 mph.

In North Carolina, no super hurricanes have been verified, although extreme winds have been recorded in several storms. On August 18, 1879, winds at Cape Lookout were estimated at 168 mph, after the lighthouse keeper's *anemometer* (wind-measuring instrument) was blown away. Several estimates of 150-mph winds were reported during hurricane Hazel in 1954, and gusts of more then 100 mph were observed as Hazel tracked northward through Raleigh.

The widespread damages that result from these high winds are obvious in the aftermath of a severe hurricane. Shattered, uprooted, and fallen trees are

commonplace, affecting tall pines as well as mature hardwoods. Memories still linger of hurricane Fran's impact on inland portions of the state. Massive trees were toppled by strong gusts, crushing cars and roofs and blocking streets in virtually every neighborhood along its path. As hurricane Hugo blazed across western North Carolina, its freakish winds damaged more than 2.7 million acres of forests in twenty-six counties, creating losses of more than $250 million in timber alone. And when trees fall, they often fall on power lines, homes, automobiles, and, sometimes, people. But the destruction from

On the morning after hurricane Donna's visit in September 1960, merchants along Front Street in Beaufort survey their losses. The combination of high tides and destructive winds affected almost every business along the waterfront. (Photo courtesy of Roy Hardee)

Hurricane Emily's winds easily snapped these pine trees near Cape Hatteras in 1993. (Photo courtesy of Drew Wilson/ Virginian-Pilot/Carolina Coast)

Miles from the coast, this automobile dealership in Clinton lost its sign to hurricane Hazel's winds when the storm swept inland.

Hurricane Donna's winds cracked numerous utility poles in the east, including this one near Morehead City.

Hugo in North Carolina was only a fraction of the damage left in South Carolina, where this hurricane made landfall.

Much like trees, utility poles sometimes sway or snap in a hurricane's violent gusts. As the poles go down, they bring with them a tangled web of hot electric cables, cutting power to whole communities and threatening unsuspecting storm victims with the risk of electrocution. Utility companies are always challenged by major hurricanes, as line crews work around the clock to restore power to thousands of customers after a storm has passed.

Hurricane-force winds affect structures to varying degrees, depending upon the velocity of the wind, the exposure of the building, and the materials and methods used in the construction process. Windows can be protected by shutters or plywood (tape does nothing to stop breakage). Roof failures can occur

Winds from Hazel were officially estimated near 150 mph in Brunswick County and continued at hurricane force as the storm sped across the state. Wind damage was heavy in thirty North Carolina counties. (Photo courtesy of News and Observer Publishing Co./N.C. Division of Archives and History)

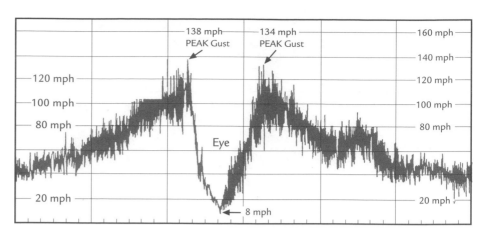

Representative anemometer record during the passing of a severe hurricane

Unsecured lumber, tools, lawn furnishings, and other loose objects can become deadly projectiles during hurricanes.

during severe hurricanes, although proper construction techniques can limit roof damage in moderate storms. But even when all building codes are followed, there is no such thing as a "hurricane-proof" house.

Fierce hurricane winds have been known to topple large airplanes, derail massive freight trains, and flip mobile homes like matchboxes. Lawn furniture, lumber, and other household items become deadly projectiles when they are launched by gusts of 100-plus mph. Those who venture out into the fury of these destructive forces place their lives in jeopardy. Cleanup efforts often take weeks or months because a storm's churning winds may leave tons of debris scattered over many miles.

STORM SURGE

Although hurricanes are often feared for their tempestuous winds, their most deadly impact comes from the accompanying onslaught of water and waves. The ocean's rapid rise peaks near the time the hurricane makes landfall, creating a *storm surge* that can be devastating. The surge is usually greatest on the beaches, but the massive rush of water that overwashes the coast can flood inland sounds and rivers far beyond their banks. And, as expected, the more intense the storm, the greater the storm surge.

Our first sign of the storm surge's approach may arrive a day or more before the hurricane, as the bulging ocean begins to pile up several feet of water against hundreds of miles of coastline. Huge swells travel great distances in front of the storm, crashing on the beach and creating a roar that can be heard for miles inland. Although the skies may be clear and sunny, the ominous spectacle of evenly spaced, ten-foot swells spanning the horizon gives warning of the hurricane's approach.

Unlike the single, giant tidal waves that are miscast as hurricane threats in Hollywood films, the storm surge phenomenon may inundate the coast over a period of minutes or hours. As the hurricane churns across the open sea, the combined effects of the storm's lowered barometric pressure and strong, inward-spiraling winds create a deep, swirling column of water beneath the ocean's surface. This effect causes the sea level to rise in the vicinity of the storm, creating a dome of water that may be a few feet high in the center and one hundred miles wide. This dome of water and underlying circulation advance with the hurricane and create a dangerous surge as the storm moves over shallow water near the coast.

As the hurricane thrusts toward land, the gradually shallowing seafloor forces the water dome to rise dramatically. Powerful hurricanes produce surges that can exceed heights of twenty feet above sea level, bringing total devastation to beachfront structures. Typically, pounding waves driven by high winds ride atop the measured surge. Under these extreme conditions, it is not difficult to understand why nine out of ten hurricane-related deaths are attributed to drowning.

Although some hurricanes produce monstrous storm surges, others do not. In each storm, a variety of factors contribute to the severity of the resulting flood. The hurricane's intensity and forward speed, as well as the coincidental timing of normal astronomical high tide, determine the measured effect of the tidal surge. Geographical factors such as the curvature of the coastline

Many areas of the North Carolina coast can be affected by floods, especially portions of the Outer Banks. Streets in Kitty Hawk flooded during the Halloween storm of 1991. (Photo by Drew Wilson; courtesy of the Outer Banks History Center)

Wind-driven waves sometimes overwash the primary dunes that line the shore on North Carolina's barrier islands. After this dune was eroded in the Ash Wednesday Storm, these cottages were unprotected from the advancing tide. (Photo by Aycock Brown; courtesy of the Outer Banks History Center)

and the topography of the seafloor can alter the surge's impact, by either dispersing the water dome or concentrating its destructive energy.

Historically, storm tides around the world have produced epic tragedies. In parts of Asia, the death tolls from cyclones have reached biblical proportions. As late as 1970, a disastrous storm tide claimed several hundred thousand lives in the low coastal regions of what is now Bangladesh. Eleven thousand more perished in a cyclone that struck that region in 1984. Our hemisphere has not experienced surge-related loss of life on that scale, although great tragedies have occurred. In August 1893, a powerful hurricane surprised resi-

dents of Charleston, South Carolina, and between one and two thousand persons drowned in the rapidly rising storm tide. In October of the same year, two thousand more were lost in Louisiana as another hurricane surged across the Mississippi Delta. The great Galveston hurricane of 1900 claimed eight thousand lives, most of which were lost to the spectacular rise in sea level that occurred as the storm made landfall.

Through the years, numerous hurricanes have struck the low-lying beaches of North Carolina, some completely overwashing the barrier islands that line the coast. Often, the storm tides and hurricane winds push around the waters of the Pamlico and Albemarle Sounds, swelling rivers and tidal creeks and flooding streets and homes far from the Atlantic Ocean. High winds drive surges of water across the shallow sounds, only to rebound back to sea as the winds shift, overwashing the Outer Banks and cutting new inlets across the narrow strips of sand.

Although hurricane Hazel did not sweep over the Outer Banks, its storm surge on the Brunswick County beaches was the greatest in North Carolina's recorded history. Unfortunately, Hazel's landfall coincided with normal high tide, adding to the rise in sea level at Sunset Beach, Ocean Isle, Holden Beach, and Long Beach. Across this span of some forty miles, the storm surge ranged between sixteen and eighteen feet above mean sea level. This massive flood damaged or destroyed virtually every structure along the strand and was the focal point of one of our state's greatest natural disasters. Although hurricane Fran's maximum surge was less than Hazel's overall, initial reports indicated that it ranged from eight to twelve feet. Reliable sources at Wrightsville Beach and Surf City reported that Fran's peak surge was actually higher than Hazel's in those locations.

In this dramatic before-and-after sequence, the effects of hurricane Hazel's seventeen-foot storm surge on Long Beach can be seen. More than 350 cottages lined both sides of the beach road prior to October 1954 (top); after Hazel only five remained intact. Most washed into the marsh (bottom) or completely disappeared. Miles of protective dunes were flattened by the record tide. The arrow marks where the curve in the road used to be. (Photos courtesy of the U.S. Army Corps of Engineers)

Through the years many residents on the Outer Banks have adapted to the threat of rising flood-waters. Some have even prepared their homes by drilling holes through floorboards to allow the tide to enter, thus preventing the houses from floating away. (Photo courtesy of the Coastland Times*)*

RAINFALL

The turbulent rush of the storm tide is not the only source of flooding associated with hurricanes. Torrential rains fall as spiral bands of storm clouds empty out across the land. These rains may pour for hours or days, depending on the forward speed of the hurricane. On average, six to twelve inches of rain can be expected as a hurricane passes nearby. As hurricane Floyd poured over eastern North Carolina, accumulations of more than 20 inches were reported in some areas. The Wilmington airport measured 19.06 inches during the storm and established a new 24-hour rainfall record of 15.06 inches. Rocky Mount, Tarboro, and Greenville also received copious rains, which fell on ground already saturated by previous storms.

As large hurricanes track inland, flash floods can be deadly, especially in mountainous areas. Some of the most destructive floods in U.S. history were caused by the remnants of hurricane Agnes in 1972. After coming ashore on the Gulf coast of Florida, Agnes spun across Georgia and the Carolinas, crossed over the Chesapeake Bay, and bounced back through the state of New York. Heavy rains brought extensive flooding to the eastern seaboard, as numerous rivers crested at record levels. In North Carolina, Mount Mitchell recorded eleven inches of rain, and several rivers spilled over their banks. The Yadkin River crested 14.6 feet above flood level, inundating 86,000 acres of farms and homes. Across the eastern states, 122 deaths and over $2 billion in damages were attributed to the inland floods from hurricane Agnes.

On August 17, 1955, hurricane Diane came ashore just days after hurricane Connie struck the coast. The combined effect of the two storms was severe as heavy rains added to the tidal flooding experienced in towns like Belhaven, seen here. (Photo courtesy of Roy Hardee)

Hurricane Ione, the third of the 1955 season to strike North Carolina, delivered record rainfall to much of the eastern part of the state. In New Bern, where forty city blocks were flooded, children explored the streets by boat after the storm. (Photo courtesy of New Bern–Craven County Public Library)

TORNADOES

As if bruising winds and life-threatening floods weren't enough, hurricanes sometimes breed tornadoes that can rip through populated areas with little warning. Just as large thunderheads moving across the great plains bring

Tornadoes and waterspouts sometimes develop during hurricanes. (Photo courtesy of NOAA)

twisters to the Midwest, the hurricanes' spiral rainbands sometimes spawn tornadoes. Some, known as *waterspouts*, form over water, and others touch down on land, leaving behind narrow paths of concentrated destruction. Tornado winds can easily top 200 mph, and very little can withstand their menacing force. In 1967, hurricane Beulah struck the Texas coast and established a record for tornado activity. Over 150 separate twisters were reported as the hurricane made landfall and moved inland.

STORM INTENSITY

The awesome natural forces displayed by hurricanes can sometimes leave behind an equally awesome calamity: the ruinous scene of devastated homes, businesses, farmlands, and forests. But more often than not, hurricanes strike with lesser winds and moderate tides. Not all hurricanes are created equal, and their fickle nature often brings about changes in intensity with each news update. These changes can be critically important to coastal residents and forecasters who must make judgments about warnings and evacuations. All hurricanes are dangerous, but certainly some bring a greater potential for disaster than others.

In the last few decades, meteorologists have used the Saffir-Simpson scale to rate hurricane intensity. On this relative scale of one to five, a minimal hurricane with wind speeds of 74–95 mph is considered a category-one storm. At the other extreme, category-five hurricanes are worst-case events, where wind speeds top 155 mph and storm surges can exceed eighteen feet.

Fortunately, category-five hurricanes are not common, and only two made

North Carolina's fishing piers are extremely vulnerable to hurricanes, as even modest storms can bring damages. (Photo by Drew Wilson; courtesy of the Outer Banks History Center)

Hurricane intensity is measured on the Saffir-Simpson scale, which categorizes a storm's severity and potential for destruction on a scale of one to five. A category-one hurricane may bring modest damage and a category-five storm is a catastrophic event. Hurricane Emily was a minimal category three when it brushed the Outer Banks and battered Cape Hatteras in 1993. (Photo courtesy of Drew Wilson/Virginian-Pilot/Carolina Coast)

landfall in the United States during the twentieth century. The Labor Day Storm of 1935 washed over the Florida Keys on September 3, claiming 408 lives and punishing those remote islands with winds estimated at 200 mph. Hurricane Camille rolled into Biloxi, Mississippi, in August 1969, and then its remnants washed through Tennessee, Kentucky, and Virginia. This category-five superstorm delivered a surge of twenty-five feet at Pass Christian, Mississippi, winds estimated at over 175 mph, and a death toll that exceeded 250. Camille's rain clouds emptied out across Virginia, in some areas dumping twenty-seven inches of rain in eight hours. Flash floods raced across the state, claiming 109 lives. In all, Camille was one of the most powerful and destructive hurricanes to strike the United States in the twentieth century.

In the United States, the tragedies brought about by Camille had been unmatched by modern storms until hurricane Andrew ravished South Florida in 1992. Andrew was a strong category four as it raced into southern Dade County, bearing down on the National Hurricane Center and the area surrounding Homestead Air Force Base. Winds gusted near 175 mph and entire communities were demolished. Andrew wiped out over 70,000 acres of mangrove forests, sank or destroyed over 15,000 boats, and left more than 250,000 homeless. Homestead was particularly hard hit, as tornado-like winds ripped apart houses and shattered mobile homes. After crossing the Florida Peninsula, Andrew returned to the waters of the Gulf of Mexico and regained much of its strength. Two days later, this killer storm was back, threatening the Louisiana coast with its second wind. In all, Andrew killed forty-three in Florida, claimed fifteen more lives in Louisiana, and cost this nation more than $25 billion—by far the most costly hurricane in American history.

The destructive forces of hurricane Hazel's storm surge and high winds are evident in this image of the Breakers Hotel at Wilmington Beach. Hazel ranks as a category four on the Saffir-Simpson scale. (Photo courtesy of the N.C. Division of Archives and History)

Meteorologists use barometric pressure as a primary scale for determining the intensity of hurricanes—the lower the pressure, the more intense the storm. During the Labor Day Storm of 1935, a pressure reading of 26.35 inches was made on Long Key, Florida, which stood for fifty-three years as the lowest on record in the Western Hemisphere. In 1988 hurricane Gilbert, also a category five, struck Cozumel, Mexico, with devastating force. Prior to the hurricane's landfall, reconnaissance aircraft recorded a barometric pressure of 26.13 inches, establishing a new record. A comparison of the barometric pressures of the most intense hurricanes to strike the United States through 1999 is included in the appendix of this book.

Category-five storms are rare—less than 5 percent of Atlantic hurricanes reach that level of intensity. In North Carolina, no category-five hurricanes are known to have made landfall, though Hazel was a category four. Category-three storms like Fran are more common, striking the state on average about once a decade. Category-two storms like Floyd can still bring tremendous destruction, and lesser storms also deserve respect. Any hurricane can be dangerous, changing intensity and direction as it races toward the coast.

CATEGORY	EXAMPLE	WIND AND TIDE	EFFECTS
One	Charley (1986)	Winds 74–95 mph; surge 4–5 feet above normal	No damage to building structures; most damage to unanchored mobile homes, trees, and signs. Coastal road flooding and minor pier damage.
Two	Diana (1984)	Winds 96–110 mph; surge 6–8 feet above normal	Some damage to roofing materials, doors, and windows. Considerable damage to mobile homes, trees, signs, piers, and small boats. Some coastal evacuation routes flooded.
Three	Fran (1996)	Winds 111–130 mph; surge 9–12 feet above normal	Structural damage to some buildings; mobile homes are destroyed. Coastal structures are damaged by floating debris. Substantial regional flooding extends along rivers and sounds.
Four	Hazel (1954)	Winds 131–155 mph; surge 13–18 feet above normal	Extensive structural damage with some complete roof failures. Major damage to lower floors of structures near the shore. All terrain lower than 10 feet above sea level may be flooded, requiring massive evacuation of residential areas as much as 6 miles inland.
Five	Camille (1969)	Winds greater than 155 mph; surge more than 18 feet above normal	Complete roof failures on many residences and industrial buildings. Some complete building failures. Major damage to all structures located less than 15 feet above sea level. Massive evacuation of all residents within 10 miles of the shoreline required.

Source: National Weather Service. (Note: Surge elevations may vary locally.)

OTHER FACTORS

Rating hurricanes on the Saffir-Simpson scale is important for determining their potential for destruction. But there are other factors that can contribute to the severity of a storm's impact. The orientation, forward speed, and diameter of an approaching hurricane can be as significant as its intensity. The timing of lunar tides and the geographical features of the region can also alter a storm's effects.

In the Atlantic, the *right-front quadrant* of a tropical cyclone presents the greatest danger to coastal residents. The combined effects of the storm's for-

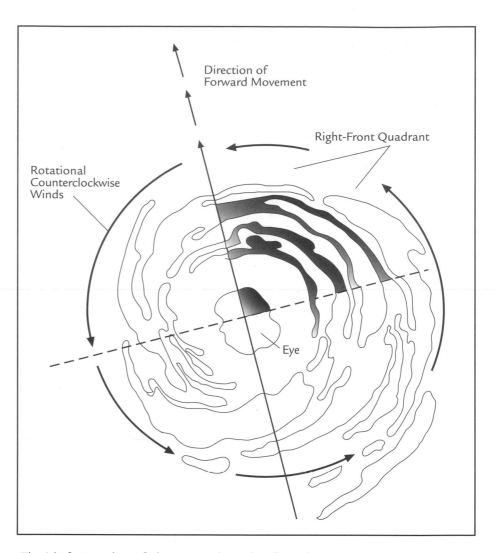

Direction of
Forward Movement

Right-Front Quadrant

Rotational
Counterclockwise
Winds

Eye

The right-front quadrant of a hurricane in the northern hemisphere

ward speed and counterclockwise rotation produce the highest winds and greatest tidal surges in areas of the coast that are hit by this portion of the storm. Typically, hurricanes approaching North Carolina arrive from the south or southeast, riding across the warm waters of the Gulf Stream. As these storms approach, their right-front quadrant is on the northeastern side. Many times, this quadrant will remain at sea as the hurricane brushes along the Outer Banks. But when a storm moves inland, the coastal region just to the right of landfall is likely to suffer the greatest damage.

The forward momentum of a hurricane as it crosses the coastline can also play a role in a storm's severity in any given area. Typically, hurricanes move across the tropical ocean at about 8–15 mph. But as they enter more temperate latitudes, they often increase forward speed, sometimes racing northward at 25–40 mph. As a storm crosses the coast, the actual measured speed of the wind could be thought of as a combination of the rotational winds of the hurricane and the contribution, either positive or negative, from the translational forward speed of the storm. Consequently, fast-moving hurricanes may bring higher winds to some locations. Slow-moving storms, on the other hand, may have lesser winds but typically dump larger amounts of rain.

Frequently, hurricanes approaching North Carolina from the south only skirt the eastern edge of the state, affecting the Outer Banks but sparing the mainland. These storms may actually make landfall near Ocracoke or Buxton while their most destructive energies remain over the Atlantic. In 1985, hurricane Gloria was just such a storm, striking Cape Hatteras and then skipping out to sea. Gloria's powerful right-front quadrant remained over the ocean while it passed, lessening the storm's destructive effects. But Gloria still ravaged the Outer Banks, causing severe beach erosion, killing one person, and causing $8 million in losses.

WATCHING THE STORMS

Amateur radio operators kept round-the-clock vigils at their stations during many of the hurricanes of the 1950s. Communications about storm movements, evacuations, and damage reports improved considerably during this period. (Photo courtesy of the Carteret County Historical Society)

Residents along the Carolina coast can expect to occasionally experience the apprehension of leaving their homes as they scramble to avoid the lashing winds and rising tides of approaching hurricanes. Costly preparations and late-night evacuations may be required of thousands of families and businesses, disrupting normal activities in numerous down-east counties. But at least today's coastal residents enjoy the benefits of forewarning. Only a few decades ago, hurricane forecasts and communications were poor, leaving some communities vulnerable. Hurricanes almost always barreled ashore without warning, often with dire consequences. Hundreds of North Carolinians have perished in rapidly rising storm tides that have left little opportunity for escape.

Before any warning systems were available, some coastal residents told the lore of a hurricane's approach. Sailors and islanders watched the skies for double moons, sundogs, and the scarlet aura of a summer sunrise. Remember the adage "Red sky at morning, sailors take warning, red sky at night, sailors delight"? Predictions of hurricanes and other catastrophes were also offered after the skies turned to night. It was once widely believed that the position and alignment of the stars and planets foretold impending hurricane tragedies.

Old-timers also monitored animal behaviors and believed some were omens of approaching storms. It's been told that shorebirds gather and livestock wander in the days preceding a hurricane. Keen observers watched the rhythm of the ocean's swells as another method of predicting incoming cyclones. But with such unreliable sources and meager communications, coastal communities stood helpless against the rapid approach of hurricanes.

During the late nineteenth century, a number of devastating hurricanes washed over the U.S. coastline, killing thousands. Frustration with these deadly storms led to the creation of the United States Weather Bureau in 1890. But it was not until the Spanish-American War of 1898 that a comprehensive hurricane-forecasting service was established. President McKinley was said to have had a greater fear of hurricanes than of attack from the Spanish navy. The warning service was extended to include shipping interests and numerous ports throughout the Caribbean, with a forecast center located in Havana, Cuba. The disastrous Galveston hurricane of 1900 shocked the nation as eight thousand died, prompting the relocation of the Weather Bureau's West Indies forecast office to Washington, D.C.

But hurricane forecasting and warning was tragically ineffective in the early part of the twentieth century. During the twenties and thirties, several deadly storms struck Florida and the East Coast, killing thousands of coastal residents. Forecasts came with short notice and often never reached remote island communities. These communities were the very ones that faced the greatest risk—the horrific flood of the storm tide.

Along the boat docks of numerous coastal villages, the task of preparing for oncoming hurricanes was well practiced during the mid-1950s. Storm warnings provided adequate time for residents to "batten down the hatches." (Photo courtesy of the Carteret County Historical Society)

Along North Carolina's barrier islands, remote villages like Ocracoke and Portsmouth were especially vulnerable. Messages between these isolated areas and the mainland were transported by boat, as direct communications were not possible. One unusual method of delivering hurricane warnings was employed in the 1940s, when small planes scanned the Outer Banks and dropped warning messages in sealed containers to the isolated residents below. After receiving the news of an approaching storm, these brave families had no time to leave their island homes, only a few hours to secure their fishing boats and prepare for high winds and raging waters.

The horrible tragedies that resulted from numerous poorly forecasted hurricanes challenged scientists to better understand these phenomena. After World War II, the steering effects of the jet stream and other mechanics of cyclone development were studied intensely. At the same time, advances in technology improved the collection of weather data and led to a better understanding of hurricanes. In 1943, the first aircraft reconnaissance flight was directed into a hurricane to gather information on wind speeds, direction, and barometric pressure. Flights of the "hurricane hunters" continue today and still provide the most accurate and timely information available to forecasters.

Reconnaissance aircraft called "hurricane hunters" began flying missions into hurricanes during the 1940s to gather information on storm positions, movements, and intensity. Today, the National Oceanic and Atmospheric Administration also uses this Gulfstream IV jet to gather data from high altitudes. (Photo courtesy of NOAA, National Hurricane Center)

Today, the National Hurricane Center in Miami, Florida, is the nerve center for our nation's hurricane warning system. Technological advances over the last several decades have transformed hurricane forecasting and warning into an accurate, timely, lifesaving service. The advent of radar, computers, weather-watching satellites, television, and regional evacuation planning have improved the system dramatically. As a result, coastal residents can tune their televisions to the unfolding drama of an approaching hurricane often days before it strikes. It is evident that this early warning system has been effective in saving lives, as losses of life from modern hurricanes have shown an overall decline in the United States.

Refinements in the forecasting and warning system continue, with the addition of Doppler radar systems, supercomputer atmospheric modeling, and more advanced local planning for evacuation and recovery. But even today, the business of predicting when and where hurricanes will strike remains a tentative one. The unpredictable nature of these fast-moving storms can still leave coastal residents with very little time to evacuate vulnerable areas. Densely populated coastal beaches and remote islands like those of the Outer Banks can swell with traffic jams as vacationers and residents scramble to escape a rapidly moving storm. And even with today's advanced warning system, those who refuse to evacuate the barrier beaches could face the same perils endured by islanders one hundred years ago.

The National Hurricane Center/ Tropical Prediction Center, located on the campus of Florida International University in Miami, is the nation's nerve center for hurricane tracking and forecasting. (Photo courtesy of NOAA, National Hurricane Center)

Hurricane Emily will be remembered for many years on the Outer Banks. (Photo courtesy of Drew Wilson/Virginian-Pilot/ Carolina Coast)

WHAT'S IN A NAME?

Popular American culture is rich with famous names. Einstein, Elvis, Oswald, and Madonna—all are household words that evoke vivid images of colorful characters. Also part of our recent past are such well-known names as Hazel, Hugo, Fran, and Floyd. For many, these names conjure up nightmarish memories of tropical terrorism. The naming of hurricanes and tropical storms began in the early 1950s, in an effort to manage storm tracking, provide historical reference, and alert and involve the general public. Often, these storms developed their own "personalities," which are forever linked to the names they received. This personification of disaster may help us better understand and cope with the emotional scars left by hurricanes.

In the eighteenth and nineteenth centuries, very few hurricanes were given names. Sometimes, extreme storms were named for the islands they devastated, the ships they sank, or the religious holidays nearest the time of their approach. The Cuba Hurricane of 1811, the Santa Ana Storm of 1825, and Racer's Storm in 1837 are some of the few early hurricanes that were given names. But most early storms are known only for the dates of their occurrences, such as the Great Storm of 1750 or the Great Wind of 1885. With so few early hurricanes identified by name, references and comparisons were somewhat difficult for those who kept historical records.

In the early days of hurricane tracking in the twentieth century, references to storms were made by their position at sea—the latitude and longitude of the latest ship report. This numerical system proved to be quite confusing, especially when several tropical disturbances were active at the same time. Eventually, military weathermen began using code names for storms in alphabetic sequence—Adam, Baker, Charlie, and so on—which made each storm easier to distinguish and track. In 1953, a system was adopted that used women's names to identify individual tropical storms and hurricanes, and that practice continued through the 1970s. In 1979, women's groups and several nations lobbied the World Meteorological Organization to change the naming system to include men's names and names of international origin. A complete list of names is now cycled with each hurricane season and then repeated every seven years. The names of significant hurricanes, like Hugo, Andrew, and Fran, are retired, never to be used again. This international system of identification has proven effective in eliminating confusion and promoting awareness in coastal communities.

EARLY NORTH CAROLINA HURRICANES, 1524–1861

Records of storms and hurricanes are widely scattered throughout the history of colonial North Carolina. These early accounts are clearly incomplete, as there were certainly numerous storms that occurred during the seventeenth, eighteenth, and nineteenth centuries for which no significant record exists. Thanks to research completed by Charles Carney, Albert Hardy, and James Stevenson of the National Weather Service, early hurricane reports have been compiled for North Carolina. Portions of these records are included here as part of a sampler of our hurricane history. But be reminded that North Carolina has been struck, on average, by one hurricane every four years over the past century. We can expect that many severe hurricanes not mentioned in this chronology have assaulted the Tar Heel state.

In the age of the New World explorations and the colonization that followed, great storms often ravaged ships and settlements. The first Europeans to explore the Carolina coast may have approached during a violent hurricane. The Italian adventurer Giovanni da Verrazano sailed into North Carolina waters in 1524 and, after enduring a storm, charted the first shoal as the Cape of Feare. Two years later, in 1526, a large Spanish expedition led by Lucas de Ayllon came to the Cape of Feare in search of gold. Some scholars believe that after wrecking his ship during a "loathsome gale," de Ayllon and his men camped in the vicinity of Bald Head Island. There they may have built a new ship, perhaps the first ever constructed by Europeans in the New World.

After receiving a royal charter to establish a colony in the New World in 1584, Englishman Sir Walter Raleigh sent out an expedition that landed on Roanoke Island. Ralph Lane was named governor of the settlement, which consisted of a fort and several homes. Two years later, Sir Francis Drake cruised into Carolina waters on his way back to England from St. Augustine. He brought his great fleet of twenty-three ships to the struggling Roanoke colony to replenish badly needed supplies and to offer Governor Lane a new ship. On June 13, 1586, a four-day hurricane scattered Drake's fleet and wrecked many of his ships. Lane later reported that "in the terrible storm he [Drake] had undergone more dangers from shipwreck in his desire to bring aid to us than all his previous engagements with the Spaniards." Lane and Drake later returned to England, bringing to an abrupt end the colony known as the Roanoke Hundred.

Still determined to establish a colony in the New World, Raleigh dispatched another group of colonists to Roanoke in 1587. This new group was led by Captain John White, whose granddaughter was Virginia Dare, the first English child born in America. This second attempt to settle on the Carolina coast was also a failure. Governor White sailed to England for desperately needed supplies and returned to the Outer Banks in 1590. Upon his arrival on Roanoke Island, the colonists could not be found and their homes had

(Page 33)
The Diamond Shoals Lightship founders in a storm off the coast of Hatteras. The area, notorious for its dangerous shoals, became known as the Graveyard of the Atlantic. (Engraving from the Illustrated News, *April 23, 1853; courtesy of the North Carolina Collection, University of North Carolina at Chapel Hill)*

vanished. White discovered only a wooden palisade and the word "Croatoan" carved in a nearby tree. Some have speculated that the fate of the Lost Colony of Roanoke may be linked to a hurricane disaster. Contributing to the theory, in which a great storm surge swept through the settlement, four cannon, iron bars, and other metal debris were found around the site. Heavy objects may have been left behind while homes and ships were scattered by the tides. This theory is only one of many that have been offered on the mystery of the lost colony.

Throughout the next two centuries, coastal settlements slowly developed in the East, and with the exception of Native American peoples, the region was very sparsely populated. Few records or reports exist of the many hurricanes that must have affected the early colonial settlements. Through the seventeenth century, only three hurricanes are known to have affected North Carolina, although many others must have struck the region. On September 6, 1667, a severe storm dragged through southern Virginia destroying crops and buildings. It is assumed that this hurricane passed over the Outer Banks prior to its move up the Chesapeake. Rains from this storm were said to have lasted for twelve days.

The Great Storm of August 18, 1750, was responsible for numerous reports of damage along the Carolina coast. New inlets were cut through barrier islands, and five ships of the Spanish Flota, a fleet sent to plunder coastal settlements, were reported washed ashore or wrecked. Then, in 1752, another fateful storm tracked from Charleston, South Carolina, up the coastline and also destroyed many ships. But this hurricane's most infamous result was the flood and destruction of the Onslow County seat. The town of Johnston, named after Governor Gabriel Johnston, was built on a bluff in an area now known as Old Town Point, part of present-day Camp Lejeune. During this late September storm, the Onslow County Courthouse was completely destroyed, and all of the county's records and deeds were lost. Portions of the courthouse were said to have washed "across the New River, there two miles wide." Virtually every building in town was reportedly wrecked, and eight residents were killed. So great was the loss at Johnston that the town was abandoned and a new county seat was established at Wantland's Ferry (known today as Jacksonville). The only structure that remained at Johnston was the jail, as reported in court documents in 1753: "Whereas the prison is not sufficient since the storm and as no one lives near it, it is the opinion of the court that the sheriff may make a prison of his own house or plantation till further provision is made."

According to legend, rising water from the hurricane of 1752 swept a small boy, about four years old, across the river, where he was saved from the deadly tide. Barely able to speak after his frightening ordeal, the only word he could say was "Hadnot." The point of land where he was found was then named Hadnot Point, which is also part of present-day Camp Lejeune. The boy's name was Charles Hadnot, and he was adopted by the county.

A severe hurricane in September 1761 washed over the southern coast of North Carolina. Once again many ships were wrecked and homes were destroyed. A new inlet was cut near Bald Head Island at a location known as "Haul-over." Nearly one mile wide and eighteen feet deep, this inlet remained open for more than a hundred years.

Lengthy reports were made of a devastating hurricane that struck North Carolina on September 6, 1769. The effects of this storm were most severe in the region from Smithville (known today as Southport) through New Bern, the colonial state capital. The Brunswick County Courthouse was reportedly blown down, along with thousands of trees. In New Bern, the tide was said to have risen twelve feet higher than "ever before." Many homes and stores were destroyed, and one entire street of houses was washed away, along with several residents. In a letter to the Earl of Hillsborough, Governor Tryon wrote: "New Bern is really now a spectacle, her streets full of the tops of houses, timber, shingles, dry goods, barrels and hogsheads, empty most of them, rubbish, . . . in so much that you can hardly pass along; a few days ago so flourishing and thriving—it shows the instability of all sublunary things. . . . In short, my Lord, the inhabitants never knew so violent a storm; every herbage in the gardens had their leaves cut off. This hurricane is attributed to the effect of a blazing planet or star that was seen both from New Bern and here, rising in the east for several nights between the 26th and 31st of August, its stream was very long and stretched upwards towards the west."

Notable hurricanes struck the New Bern area again in 1803, 1815, 1821, and 1825. On September 3 and 4, 1815, a powerful hurricane surprised coastal residents and made landfall near Swansboro. According to a report in the *Raleigh Minerva*, the storm caused great damage and loss of life in Onslow. Mr. Nelson's home on Brown's Banks was swept away during the storm surge, taking with it four of his children. The father and one son survived by clinging to the wreck of their house as it carried them nearly twelve miles to Stone's Bay on the New River.

A rare early June hurricane swept through North Carolina in 1825, leaving destruction in its wake from Cuba to New England. Tides at Adam's Creek rose fourteen feet, and surging water flooded downtown New Bern. More than twenty ships were driven ashore on Ocracoke Island, twenty-seven near Washington, and dozens more from Wilmington to Cape Lookout.

On August 24 and 25, 1827, another powerful hurricane moved across the state, with reported effects from Cape Hatteras to Winston-Salem. During the peak of this storm, the Diamond Shoals Lightship broke away from its anchors and drifted southward to Portsmouth. Two of the lightship's crew were washed overboard and lost at sea. After this storm, the treacherous Diamond Shoals were without a signal light for several years.

The year 1837 was significant in North Carolina's hurricane history, as three storms are known to have struck the state between August and November. The hurricane that inundated the coastal region on August 19 of that year most likely came ashore near Wilmington. This storm brought tremendous rains to the region, and rivers crested at record levels. It was reported that there was not a bridge left standing between Wilmington and Waynesboro (known today as Goldsboro). According to one eyewitness report from the storm, "The gale was certainly the most violent we have witnessed and the quantity of water . . . greater than has ever been known."

In October of the same year, a long-lived hurricane dubbed Racer's Storm wandered from the Yucatan Peninsula to the Texas coast, across the Gulf states and Florida and into the Atlantic. On October 9, 1837, this storm crossed over the Outer Banks, sinking numerous ships. One of this hurricane's worst tragedies occurred with the loss of the steamship *Home*, as 90 of the ship's 130 passengers were lost at sea. Three weeks after Racer's Storm had passed, the third hurricane of the season bruised the Outer Banks.

In 1842, two hurricanes punished coastal North Carolina. Damage was recorded from Wilmington to Currituck after the first, a July storm that sunk numerous ships along the coastline. The most severe damage was reported from Portsmouth northward along the Outer Banks, where livestock drowned and homes were washed away. For residents of these remote islands, this hurricane is believed to have been one of the most severe on record. Less than thirty days later, another hurricane swept over the Outer Banks, again bringing destruction to the region. Among the losses were three ships: the *Congress*, which wrecked at Cape Hatteras, the *Pioneer* at Ocracoke, and the *Kilgore* at Currituck.

The hurricane that approached the North Carolina coast on September 6, 1846, was both intense and slow moving. A remarkable surge of water, driven by continuous northeast winds, pushed far into the Pamlico and Albemarle Sounds, flooding rivers and creeks for miles inland. Then, as the hurricane passed and its winds rotated to the southwest, this massive expanse of water rushed back toward the sea, overwashing the Outer Banks from west to east. On the night of September 7, a new inlet was created by these events, known today as Hatteras Inlet. The next day, a second inlet was formed just south of Roanoke Island. This inlet soon became navigable and was named Oregon Inlet for the first large boat to pass through it, the *Oregon*. For years after this storm, sounds and bays that had always been freshwater were said to contain oysters, stingrays, and other saltwater creatures. In addition to wrecking homes and ships, this amazing hurricane literally reshaped the geography of the Outer Banks.

The "perfect tempest" that struck the Cape Fear region on a full moon in September 1856 also delivered a massive storm surge. Heavy crop damage was reported, as fields were flooded with salt water from the hurricane's incredible tide. Prior to this storm, Wrightsville Beach was said to have been covered with groves of live oaks. As this hurricane made landfall, the surging ocean overwashed Wrightsville. The waves uprooted and swept away most of the oaks and left only a few trees standing. Of those that remained, most died within a few days due to the invasion of salt water. Reports of waves breaking one-half mile inland from the sound at an elevation of thirty feet have led to speculation that this hurricane's floods may have been some of the worst in North Carolina's history.

Just after the battle of Fort Sumter in April 1861, President Lincoln declared a commercial and military blockade of all southern ports, and the Civil War began. Within months, the Union navy positioned a fleet of seventy-five vessels along the Carolina coast, which was at that time the largest ever assembled by a U.S. commander. On November 1, a terrific late-season hurricane scattered the fleet and brought a major setback to the Union command. At least two vessels sank, and several sailors drowned. Some of the ships were wrecked on North Carolina beaches, where Confederates were able to salvage their goods.

Weather station at Cape Hatteras, ca. 1890. (Photo courtesy of the Outer Banks History Center)

Coastal North Carolina enjoyed a prosperous period through the late nineteenth century, as many cities and towns became linked by waterway, railroad, and telegraph. Growing agricultural and fishing industries provided work for residents, and the lure of the ocean attracted an increasing number of Piedmont vacationers. The coastal population was growing at a steady pace, even though great storms frequently battered eastern communities.

The twenty-five-year period that ended the century was a particularly active one for hurricanes in North Carolina. At least eleven struck the state during that time, and five of those were severe. On several occasions, two or more struck the coast during the same year, sometimes only weeks apart. This kind of double whammy occurred in 1882, 1893, and again in 1899.

The recording and reporting of information about hurricanes improved significantly during this period. Weather stations positioned across the state made records of winds and events, building an overall picture of each storm's movement and character. Improvements in communications brought more accurate and timely reports of hurricane damages to newspapers in the East. As a result, we know many more details about the hurricanes of the late nineteenth century than those of earlier times.

SEPTEMBER 17, 1876

Early on the morning of September 17, 1876, a powerful hurricane inundated the Cape Fear region, bringing significant damage to Smithville, Brunswick,

and Wilmington. High winds disabled the recently installed anemometers at
Wilmington and Cape Hatteras, leaving no good record of the storm's top
winds. In Wilmington, this hurricane was called "the worst in many years,"
and water rose "unprecedentedly" high in the sounds. Many large trees were
downed, bridges were washed away, and two railroad box cars were said to
have been driven uphill by the wind. Waterfowl were forced inland, and marsh
hens were said to have been killed with sticks as they sought refuge in homes
and barns. By nine o'clock that morning, the skies over Smithville (Southport) were "as dark as any night" as the storm's heavy rain clouds spun their
way northward.

The impact of the 1876 storm was felt far beyond the Cape Fear region. In
Onslow County, a military camp established to install and maintain the area's
first telegraph line was destroyed by the rising tide, and two soldiers drowned.
From Ocracoke to Rocky Mount, reports were gathered of killed and injured
citizens. Although this storm brought many tragedies to the state, it also is
credited with bringing a minor miracle to Hyde County.

It seems that in the town of Swan Quarter, in the spring of 1876, the local
Methodists had decided to build a new church. After selecting a desirable location near the center of town, the congregation was displeased to learn that
the land's owner, Sam Sadler, had no interest in giving up his property. Even
after the offer was increased, Sadler refused to sell. Determined to construct a
new church, the citizens obtained another piece of property on the edge of
town, where they built a small frame building. The congregation was satisfied,

and their new church was dedicated on September 14—the same day a major hurricane was churning past Cuba on its way toward the Carolina coast.

As the hurricane spun across the state, winds drove high waters across Pamlico Sound and piled them on the shores of Hyde County. Swan Quarter was flooded with five feet of water. Homes and businesses were deluged and wrecked, and the town's fishing fleet was severely damaged. But even with all the destruction around them, the residents of Swan Quarter were most alarmed by an apparent act of divine intervention.

During the storm, rising tides in the street had lifted the small frame church off its foundation and floated it toward the center of town. After the waters receded, residents were astonished to see that the new church had settled down on Sam Sadler's land, just as they had originally planned. Sadler was also impressed: he later signed a deed and gave his land to the Methodist church. Today, a sign stands in front of the Providence Church, reminding visitors that this was the church "Moved by the Hand of God."

AUGUST 18, 1879

In the late nineteenth century, seaside resorts in coastal North Carolina were frequented by vacationers from across the state, just as they are today. One of the more popular resort destinations was the prestigious Atlantic Hotel in Beaufort. Built in 1859 by Captain Josiah Pender, the hotel was constructed on pilings over the water at the foot of Pollock Street. During the Civil War the resort was transformed into a hospital, but by 1870 the Atlantic Hotel was again hosting vacationers. The three-story structure featured broad verandas on each level and numerous windows to catch the prevailing summer breeze off the water. Its charm and location attracted business leaders and families who traveled by train from as far west as Asheville. The hotel did not endure for long, however, as it became the scene of a tragic episode of North Carolina's hurricane history.

On August 17, 1879, preparations were under way for the arrival of a special convention in Beaufort. The *Raleigh Observer* on August 16 announced: "Major Perry of the Atlantic Hotel, Beaufort, will give a Grand Dress Ball in honor of the North Carolina Press Association on Thursday night, August 21. . . . Major Perry will spare no pains in making it one of the handsomest of the season." The hotel was already brimming with guests, including Governor Thomas J. Jarvis, his wife, and numerous prominent friends from around the state. The Gatlings of Raleigh, the Stronachs of Wilson, and the Hughes of New Bern joined the governor and his family. On the evening of Sunday the seventeenth, the hotel guests retired with no knowledge of the powerful hurricane that was soon to strike.

By 1:00 A.M. on August 18, heavy rains and gusting winds were sweeping over the North Carolina coastline. By 3:00 A.M., many Beaufort residents were pacing in their homes as the winds increased dramatically and the storm's surge began flooding Front Street. The hurricane was intense at Cape Lookout, where a Signal Corps officer reported:

> The howling of the wind and the rushing of the water past the station woke us at 5 A.M., 18th. Velocity at this time being 80 mph. and rapidly increasing. The rain pouring down in torrents, the sea rushing past the house at a fearful rate and rising rapidly. It soon undermined the Signal Service Stable, The Light House Establishment Store House and a cookhouse, which were blown down and carried away by the rushing tide. The Signal Service mule which became loose when the stable washed away tried to come to the dwelling house but could not face the raging storm; she turned and rushed into the foaming billows. The fence around the lighthouse next went carrying the keeper's fuel along with it. The whaling schooner *Seychell* of Provincetown, Mass., 50 tons, Capt. Cook, fishing in these waters, was at anchor in the Hook, parted her chains. . . . At the time the vessel crossed Wreck Point she was drawing 12 feet of water, thus showing the tide to have been fuller than ever known at this place.

Amazingly, the signal officers at Cape Lookout were able to survive the storm and were witness to the highest winds ever reported in North Carolina. The station's anemometer cups were blown away at 6:35 A.M., at which time the register showed a velocity of 138 mph. The tides continued rising and the winds steadily climbed until 7:35 A.M., when an estimate was made of 168 mph. Anemometers were also reported destroyed at Fort Macon, Portsmouth, Hatteras, Kitty Hawk, and Cape Henry, Virginia.

In Beaufort, the guests at the Atlantic Hotel were stirred by the storm in the early morning hours. One visitor reported: "About 4 o'clock A.M. the tide had risen very much, and the wind was so strong that it was impossible to stand. But those who had witnessed repeated storms at Beaufort told us there was nothing to fear. At about 5 o'clock, the water had risen to the floor of the hotel, and it was thought best to remove the children and ladies. This determination was taken so late that many of them had hardly time to dress, and a few were not dressed."

Many of the hotel guests escaped with a few clothes or belongings and were forced to swim or wade through the chest-deep waters that had engulfed Front Street. The surging tide washed boats and debris as far into town as Broad Street. Most of the hotel refugees were taken in by the good people of Beaufort, where they were sheltered in hallways and kitchens.

As the hurricane's surge continued to rise, the Atlantic Hotel and other waterfront dwellings began to crumble. In the chaos that followed, the Beaufort waterfront was the scene of heroic rescues and great tragedies. The entire

Stronach family was saved from the collapsing hotel by a black man, Palmer Davis. Davis carried the Stronach children in his arms through chest-deep water and falling debris. He also rescued the teenaged daughter of Seaton Gales of Raleigh. Davis was later recognized by Governor Jarvis as the hero of the hour, although there were others who risked their lives in the dreadful storm.

The Hester brothers from Morehead City were also credited with saving lives as they helped several guests to safety. Henry Congleton, a local boat-hand, drowned in the tide as he attempted to rescue desperate vacationers from the top floor of the hotel. Two young men from New Bern, Owen Guion and Justice Disosway, were among those trapped on the hotel's third floor. They apparently made a last-minute escape when they jumped into the rising water, clinging to their mattresses with money in their mouths.

In addition to Congleton, two other men drowned in the collapse of the Atlantic Hotel: John Dunn of New Bern, a guest at the time, and John D. Hughes, a local young man who was one of the first to offer assistance to frightened vacationers. After rescuing several guests from the second and third floors, Hughes returned to the hotel during the peak of the storm. Thinking he saw a young woman in a window, he again approached the structure, just as battering waves and wind caused the hotel to collapse. His death was in vain, however, as he had apparently mistaken the window's shimmering curtains for the nightgown of a woman.

Among those who escaped the hotel with no time to spare were Governor Jarvis and his wife. It's been told that the governor, like many other survivors, lost all of his personal belongings and was forced to flee in his pajamas with his shoes in his hand. In the chaos of the escape, his shoes were lost, but he managed to lead his family to the safety of a cottage several blocks away.

After the storm passed and the tide receded, the Beaufort waterfront was piled high with the wreckage of the night. Trunks of damaged goods littered Front Street along with lumber and broken skiffs. Crowds of dazed people, many of them barefooted, sifted through the rubble in search of lost belongings. Thousands of dollars worth of jewels were reportedly lost in the destruction of the Atlantic Hotel. Governor Jarvis called out fifty men from the garrison at Fort Macon to guard the property strewn about the waterfront.

The citizens of Beaufort took in more than 150 refugees and offered them clothes, shoes, and whatever food was available. Governor Jarvis, in need of proper clothing, was given a sailor suit that had last been worn in the War of 1812. No shoes could be found that would fit his large feet, so he wore a pair of oversized boots. Mrs. Jarvis borrowed a calico housecoat and was thankful to be dressed for the trip back to Raleigh.

On the morning of August 19, the refugees were transported by boat to Morehead City, where they boarded a train to take them home. At New Bern, a large crowd had gathered at the station to express sympathy to Major and Mrs. Hughes for the tragic loss of their son. At 8:30 P.M. the train arrived at the station in Goldsboro, where forty editors of the North Carolina Press Association had gathered to meet the survivors. The press association had canceled their trip to the coast upon news of the hurricane and instead were meeting at Goldsboro's Gregory Hotel. The newsmen would later report on the weariness of the group and the tragedies of the storm.

Upon their arrival in Raleigh, Governor Jarvis encouraged the remaining survivors to gather at the Yarborough House Hotel for refreshments. There the storm-battered group made a toast to their survival, and the reception lasted into the night. One newsman reported that the governor looked "as weather-beaten as he used to after one of Lee's campaigns." That evening discussions about rebuilding the Atlantic Hotel began. Apparently, the idea was modified, as the New Atlantic Hotel was not built in Beaufort but in Morehead City, where it became a vacation landmark for many years.

The devastating effects of the great hurricane of 1879 were felt far beyond the Beaufort waterfront. The storm apparently made landfall near Wilmington, crossed the Pamlico Sound, and returned to the Atlantic near Norfolk, Virginia. From there it recurved to the north and brought record tides to Atlantic City and Boston. Dozens of ships were wrecked from Smithville (Southport) to Cape Cod. But by far, the storm's greatest effects were evident in eastern Carteret County.

In Morehead City, the losses were heavy and included one thousand feet of railroad track, the Morehead Market House, several windmills, a Methodist church, the city wharves, and dozens of shops and homes. Virtually every structure lost its chimney in the 150-mph gusts. Thomas Webb, a night watchman at the railway depot, nearly drowned when the rapidly rising water separated him from the mainland. He saved himself by tying his body to a drifting platform with his pants. When he was later found, exhausted, he had lost most of his clothes in the raging wind and water.

The hurricane's storm surge opened at least two inlets on Bogue Banks, just west of Fort Macon. Great destruction was reported on the barrier island communities of Diamond City and Portsmouth. Beaufort Inlet was reshaped, as almost eight hundred yards of sand were washed away on the western end of Shackleford Banks. Twenty-one dwellings were leveled in the town of Smyrna, and other communities suffered great destruction, including Cedar Island, Ocracoke, Hatteras, and Kitty Hawk.

Stories of the great hurricane of 1879 have been passed down to the families of those who survived it. Much was written about this storm's severity and destruction, and some labeled it the "worst ever." But labels can be deceiving, as each generation may endure a tragic hurricane and then describe it as incomparable. Other hurricanes that struck the North Carolina coast may have had greater winds or higher tides, but to the seaside vacationers in 1879, this storm was the worst.

SEPTEMBER 9, 1881

On September 9, 1881, a severe hurricane struck near Smithville (Southport) and curved northward through Wilmington on a path through Norfolk, Virginia. Witnesses in Smithville reported this storm to be the most violent in fifty years, as the town was "covered with fallen trees, scattered fences, and the debris of demolished buildings. All pilot boats in the harbor were sunk, and loaded vessels driven ashore."

At Wrightsville Beach, the hurricane passed around noon, its winds changing direction from the east upon approach to the west after passing. These western winds brought the greatest destruction, "blowing with redoubled fury, crushing buildings and tearing up the largest trees." At Wilmington, the anemometer was disabled after the recorder measured a four-minute constant wind of 90 mph. Property damage in the Wilmington area was estimated to be $100,000.

1882

The 1882 hurricane season was active all across North Carolina. A great hurricane came ashore on the Gulf Coast on September 9, washed through Georgia and South Carolina, and moved into the Tar Heel state near Charlotte on the eleventh. After crossing to the east, this powerful storm returned to sea near the Chesapeake Bay. Great rains fell and violent winds blew, affecting the Carolina coastline from Cape Fear to Currituck. Fences, chimneys, and large trees were toppled across the East.

Less than two weeks later, another hurricane struck the state. A cyclone crossed Cape Lookout on September 22, bringing heavy rains to the eastern counties while tracking through the Pamlico and Albemarle Sounds on its way to Virginia. Swollen rivers washed out several bridges, including a trestle on the Wilmington and Weldon Railway. One train reportedly crashed in the washout, seriously injuring the crew aboard. Crop damage from the battering rains was heavy. Tarboro reported the heaviest rainfall since 1842, and some areas reported almost eight inches within a few hours.

Three weeks later, the third tropical weather event of the season brushed past the Carolina coast. On October 11, heavy rains fell throughout the day in Wilmington, as the storm apparently remained at sea. Damage from this storm was minimal along the southern coast.

SEPTEMBER 11, 1883

In September 1883, a violent hurricane made landfall near Smithville and brought punishing winds to the Cape Fear region. On the morning of the eleventh, maximum winds in Southport were recorded at 93 mph, with gusts over 110. Newspaper reports described a constant gale of over 80 mph that lasted for more than seven hours. Even in the inland portions of Brunswick County, trees and foliage appeared frostbitten after the storm because of the salt spray carried off the ocean by the winds. Trees, fences, and telegraph lines were downed, and there was severe crop damage as far inland as Harnett County. Wind-driven water pushed far up the Cape Fear River, flooding large portions of its western banks. The Hotel Brunswick in Smithville served as a shelter for the women and children of the area.

Among the losses were several pilot boats and other craft, many of which broke their moorings and were scattered by the storm. The Frying Pan Shoals Lightship was torn from its anchors and came ashore near Myrtle Grove Sound. Countless other schooners and barks were either grounded or sunk. The thirty-three passengers aboard the steamer *City of Atlanta* survived a frightening ordeal at sea during the hurricane. Battered, but with all accounted for, the disabled steamship was towed into Smithville after the storm. Several drownings were reported from other wrecked ships, and many others died in homes that either were flooded or had collapsed during the violent gale. In all, fifty-three North Carolinians are known to have lost their lives in this hurricane, more than in any storm in the state's history.

AUGUST 25, 1885

On August 24, 1885, a powerful hurricane raced through the Bahamas and made landfall near Savannah, Georgia. As it curved to the northeast, this storm washed over the South Carolina coast, claiming twenty-one lives on its way north. On September 25, the hurricane passed just west of Wilmington and continued its northeastern arc toward Cape Hatteras. At 5:15 P.M. the anemometer in Smithville was blown away after a reading of 98 mph. Within the next half hour, winds were estimated to have surpassed 125 mph at that location.

Destruction from this cyclone was reported along the southern coast and from inland counties as well. Crop destruction was heavy and many ships were grounded. Damage in Smithville was estimated to have exceeded $100,000, but the destruction in Charleston, South Carolina, approached $1.7 million. After the heavy losses on the Atlantic coast from the 1885 storm, a weather-reporting network was proposed for the West Indies and Mexico.

AUGUST 27–29, 1893

The year 1893 was deadly for hurricanes. Of the three powerful cyclones that struck the United States between August and October, two brought catastrophic destruction and great loss of life to the Carolinas.

The Great Hurricane of 1893 caused horrible flooding as it came ashore near Hilton Head, South Carolina, and totally inundated the low-lying communities of the South Carolina coast. Between one and two thousand people were believed to have drowned as this hurricane's massive storm surge surprised residents and reshaped the coastal islands. Sometimes called the Sea Islands Hurricane of 1893, this great disaster left more than thirty thousand homeless. For more than nine months, many of the storm's refugees survived on a ration of "a peck of grits and a pound of pork, per family, per week."

The Sea Islands Hurricane charted a course that would be followed by hurricane Hugo almost one century later. After killing thousands at the coast, the storm moved inland, passing through Charlotte before curving to the northeast. The massive size of this hurricane brought 72-mph winds to Wilmington, even though the more powerful winds remained near the storm's center, almost 150 miles inland. A newspaper in Kernersville reported: "A terrific cyclone struck here at five o'clock this morning. A hundred houses wrecked and a woman killed. Many were injured. Factories, stores and residences were unroofed and some were blown away."

Along the coast, the northern edges of the storm surge pushed into the Cape Fear region. At Wilmington, "the river tide was the highest ever known here. All the wharves being submerged, a number of vessels were wrecked on the coast." As the hurricane spun across the state and into Virginia, heavy rains of three to eight inches were reported at virtually all North Carolina stations.

OCTOBER 13, 1893

Six weeks after the Sea Islands Hurricane, another great hurricane entered North Carolina from the south. On October 13, this storm came ashore near Myrtle Beach and beat a path through Raleigh, a pattern similar to the track of hurricane Hazel, which would follow in 1954.

Once again, flooding on the southern coast was of record proportions. Along the Wilmington waterfront, wharves that had been submerged in the previous storm were washed away and ruined. The overflowing tide in downtown Wilmington was reported to be the greatest ever, measuring sixteen inches higher than the previous high mark established in 1853.

Crop damage from both of the 1893 hurricanes was severe. Many homes, farms, and businesses were still recovering from the August storm when the

During the late nineteenth century, many gallant rescues were performed by the crews of lifesaving stations along North Carolina's Outer Banks. (Photo courtesy of the Outer Banks History Center)

October hurricane carved a path through the state. Twenty-two North Carolinians died in the later storm, and the death toll for the two hurricanes was near forty.

AUGUST 16–18, 1899

Six years after the 1893 season, North Carolina was again ravaged by two hurricanes in the same year. And, once again, these great storms made landfall in August and October. This time, however, both hurricanes made direct hits on the North Carolina coastline: one across the Outer Banks and the other just below Wilmington.

The Great Hurricane of August 1899 is often referred to as San Ciriaco and was one of the most powerful cyclones to move through the western Atlantic in the nineteenth century. It was named by the people of Puerto Rico, where it crossed without warning on August 8, killing hundreds. The following day, the hurricane swept over the Dominican Republic and then brushed northern Cuba on the tenth. Its north-westward movement brought it near Florida's prized oceanfront resorts, and on August 13, the gently curving storm swept past the Fort Lauderdale region. As it followed the warm waters of the Gulf Stream, its continued movement might have carried it east of Cape Hatteras and out of harm's way. But on the morning of August 16, its forward speed slowed considerably, its direction changed to the northwest, and it increased in strength as it moved toward Cape Lookout.

On the morning of August 17, 1899, San Ciriaco swept over the lower banks near Diamond City. Reports of great destruction from Beaufort to Nags Head were later printed in newspapers across the country. In Carteret County, the island communities of Shackleford Banks, Diamond City, and Portsmouth were especially hard hit. These fishing villages were settled by hardy families who were accustomed to foul weather and remote lifestyles. But numerous hurricanes and northeasters near the end of the century had tested the endurance of the people known as "Ca'e Bankers." These storms left drifts of barren sand that replaced the rich soils of their gardens, and saltwater overwash killed trees and contaminated drinking wells. These communities had begun to see a decline in population prior to 1899, largely due to the unwelcome effects of hurricanes.

For the residents of Diamond City and Shackleford, the San Ciriaco hurricane was the final blow. Few if any of the homes in these island villages escaped the rushing storm tide that swept over the banks. First, the waters rose from the soundside, as northeast winds pounded the islands during the hurricane's approach. Then, as the storm passed, the winds shifted hard to the southwest, surging the ocean's tide over the dunes until the waters met. Cows, pigs, and chickens drowned, all fishing equipment was destroyed, and many homes were ruined. The aftermath was a truly ghastly scene, as battered caskets and bones lay scattered, unearthed by the hurricane's menacing storm surge.

Following the San Ciriaco storm, the people of Diamond City and Shackleford Banks gathered their remaining belongings and searched for new places to live. Many moved to the mainland, settling in Marshallberg, Broad Creek, and the Promised Land section of Morehead City. Others moved down to the island of Bogue Banks and became squatters among the dunes of Salter Path. But most chose to relocate within sight of their former community, three miles across the sound on Harker's Island. Some even salvaged their island homes, floating them across the water on barges and repositioning them on new foundations.

One of the great tragedies of the hurricane of August 1899 fell upon several families from down-east Carteret County. August was mullet fishing time, and a large group of men gathered their nets, tents, and provisions for a two-week expedition to Swan Island, just as they did each summer. Their means of transportation was a small dead-rise skiff, twenty-one feet long and about five feet wide. Each shallow skiff could carry two men and their equipment, and each craft featured a small sail on a removable mast. These shallow draft boats provided effective transport around the protected waters of Core Sound.

This particular August, the group of twenty fishermen had already established their camp on the remote island when the first signs of the San Ciriaco
hurricane were recognized. At first, the brisk winds and gathering clouds appeared to be just a good "mullet blow," which would get the fish moving. But
on the morning of August 17, the tide was unusually high, and heavy rains
began to sweep through the sound. Alarmed by the rising water, the fishermen considered leaving but chose to stay on the island for fear of the ever-
increasing winds. They were forced to pack all of their nets and supplies aboard
their skiffs, as the tides washed completely over the island. They moored their
skiffs as close together as they could and crouched under their canvas sails for
protection from the driving rain. This proved useless, however, as they soon
had to bail the water that rapidly filled their boats.

The fishermen worked frantically to keep their skiffs afloat while 100-mph
winds churned the waters and tested their anchor lines. For several hours, the
courageous men rode out the storm, until finally, in the early hours of August
18, the winds subsided. The tide was now unusually low, as the hurricane's
winds had pushed a surge of water westward up the Neuse River. Battered but
still together, the fishermen debated making a run for the mainland, as they
could now put up sail. They knew that this journey of less than ten miles
would test their skills. Not all agreed to the plan, but after a few had left, the
others soon followed. This proved to be a great mistake. The lull that gave
them the opportunity to leave was nothing more than the passing of the hurricane's eye over Swan Island. Within minutes, the storm's winds were again

full force, this time gusting from the southwest. The small skiffs were now out on West Bay, and most were capsized by the wind and waves when a ten-foot surge of water washed back from the Neuse River.

Only six of the twenty men who left the island survived. Among those who were rescued were Allen and Almon Hamilton, who saved themselves by quickly taking down their mast and sail, throwing their nets overboard, and lying low in their skiff as it was tossed about. Fourteen others were not as lucky. Of those who drowned, ten were from Sea Level: Joseph and John Lewis, Henry and James Willis, Bart Salter, John Styron, William Salter, John and Joseph Salter, and Micajah Rose. Four brothers from the community of Stacy were lost: John, Kilby, Elijah, and Wallace Smith.

Ocracoke Island was also hard hit by San Ciriaco. The August 21 edition of the *Washington Gazette* reported: "The whole island of Ocracoke is a complete wreck as a result of the fierce storm which swept the entire coast of North Carolina, leaving ruin and disaster in its path. . . . Thirty-three homes were destroyed and two churches were wrecked. Practically every house on the island was damaged to some extent." The article also reported that waves twenty to thirty feet high pounded the beach and that the hurricane's storm tide covered the island with four to five feet of water. Hundreds of banker ponies, sheep, and cows drowned. The dazed survivors of Ocracoke endured "much suffering" after the storm from a lack of food and water.

The residents of Ocracoke and other Outer Banks communities were wise to the effects of rising hurricane tides. Many installed "trap doors" in the floors of their homes to allow rising water to enter, thus preventing the structure from floating off of its foundation and drifting away. Some simply bored holes in the floorboards to relieve the water's pressure. Occasionally, desperate times called for desperate measures. The late Big Ike O'Neal described his adventure in the '99 storm to Associated Press columnist Hal Boyle: "The tides were rising fast and my ole dad, fearful that our house would wash from its foundations, said 'Here son, take this axe and scuttle the floor.' I began chopping away and finally knocked a hole in the floor. Like a big fountain the water gushed in and hit the ceiling and on top of the gusher was a mallard duck that had gotten under our house as the tides pushed upwards."

Hatteras Island was devastated by the August hurricane of '99. The Weather Bureau station in Hatteras Village was hard hit, as the entire southern end of the Outer Banks fell within the powerful right-front quadrant of the storm. Winds at the station were clocked at sustained speeds of over 100 mph, and gusts were measured at between 120 and 140 mph. Ultimately, the station's anemometer was blown away, and no record was made of the storm's highest winds. The barometric pressure was reported as near twenty-six inches, which, if accurate, would suggest that the San Ciriaco hurricane may have reached category-five intensity.

One of the most chilling accounts of the storm was a report filed with the Weather Bureau office in Washington, D.C., by S. L. Dosher, Weather Bureau observer at Cape Hatteras. The following excerpt from his report details the extent of the storm surge and the struggle for survival endured by the residents of Hatteras Island:

August 21, 1899

This hurricane was, without any question, the most severe of any storm that has ever passed over this section within the memory of any person now living, and there are people here who can remember back for a period of over 75 years. I have made careful inquiry among the old inhabitants here, and they all agree, with one accord, that no storm like this has ever visited the island. Certain it is that no such storm has ever been recorded within the history of the Weather Bureau at this place. The scene here on the 17th was wild and terrifying in the extreme. By 8 A.M. on that date the entire island was covered with water blown in from the sound, and by 11 A.M. all the land was covered to a depth of from 3 to 10 feet. This tide swept over the island at a fearful rate carrying everything movable before it. There were not more than four houses on the island in which the tide did not rise to a depth of from one to four feet, and a least half of the people had to abandon their homes and property to the mercy of the wind and tide and seek the safety of their own lives with those who were fortunate enough to live on higher land.

Language is inadequate to express the conditions which prevailed all day on the 17th. The howling wind, the rushing and roaring tide and the awful sea which swept over the beach and thundered like a thousand pieces of artillery made a picture which was at once appalling and terrible and the like of which Dante's Inferno could scarcely equal. The frightened people were grouped sometimes 40 or 50 in one house, and at times one house would have to be abandoned and they would all have to wade almost beyond their depth in order to reach another. All day this gale, tide and sea continued with a fury and persistent energy that knew no abatement, and the strain on the minds of every one was something so frightful and dejecting that it cannot be expressed. In many houses families were huddled together in the upper portion of the building with the water several feet deep in the lower portion, not knowing what minute the house would either be blown down or swept away by the tide. And even those whose houses were above the water could not tell what minute the tide would rise so high that all dwellings would be swept away.

At about 8 P.M. on the 17th when the wind lulled and shifted to the east and the tide began to run off with great swiftness, causing a fall of several feet in less than a half hour, a prayer of thankfulness went up from every soul on the island, and strong men, who had held up a brave heart against the terrible strain of the past 12 hours, broke down and wept like children upon their minds being relieved of the excessive tension to which it had been subjected all through the day. Cattle, sheep, hogs and chickens were drowned by hundreds before the very eyes of the owners, who were powerless to render any assistance on account of the rushing tide. The fright of these poor animals was terrible to see, and their cries of terror when being surrounded by the water were pitiful in the extreme.

Officer Doshoz also reported on his own personal ordeal and struggle through the hurricane flood:

I live about a mile from the office building and when I went home at 8 A.M., I had to wade in water which was about waist deep. I waited until about 10:30 A.M., thinking the storm would lull, but it did not do so, and at that time I started for the office to change the wind sheet. I got about one-third of the distance and found the water about breast high, when I had to stop in a neighbor's house and rest, the strain of pushing through the water and storm having nearly exhausted my strength. I rested there until about noon when I started again and after going a short distance further I found the water up to my shoulders and still I was not half way to the office. I had to give it up again and take refuge in another neighbor's house where I had to remain until about 8 P.M. when the tide fell so that I could reach the office. I regret that I was unable to change the wind sheet so that a record of the wind could be made from the time the clock stopped running until the [anemometer] cups were blown away, but I did all that I could under the circumstances.

The San Ciriaco hurricane also affected the northern Outer Banks with high winds and storm flooding. At Nags Head, the rising waters of the Atlantic met the wind-driven waters of Albemarle Sound, flooding the entire area, even in places where the beach was one mile wide. Overwash from the storm covered many portions of the Outer Banks, destroying dozens of homes and cottages. Some of the residents of Nags Head refused to leave their homes as the storm approached, as they were confident the rising flood would soon subside. But the water kept coming, and at last some families had to be moved to safety by patrolmen from the Life-Saving Station.

In the nineteenth century, hurricanes were often compared by the number of ships they caused to be wrecked or lost at sea. Powerful storms frequently battered the North Carolina coast and earned the region its nickname: Graveyard of the Atlantic. So many vessels and sailors were lost through the years that young captains were often given special rewards for their first safe passage by the Hatteras coast.

Front Street in Beaufort after the San Ciriaco hurricane of 1899. (Photo courtesy of Charles Aquadro)

The Great Hurricane of '99 scuttled or sank numerous ships from Wilmington to the Virginia line. In his book *Graveyard of the Atlantic*, author David Stick lists seven vessels that were wrecked on the North Carolina coast during the storm: the *Aaron Reppard, Florence Randall, Lydia Willis, Fred Walton, Robert W. Dasey, Priscilla,* and *Minnie Bergen.* Also, the Diamond Shoals Lightship was driven ashore after its mooring lines were broken by the storm's mountainous seas. Six other ships were reported lost at sea without a trace: the *John C. Haynes, M. B. Millen, Albert Schultz, Elwood H. Smith, Henry B. Cleaves,* and *Charles M. Patterson.*

It is known that at least thirty-five sailors from the wrecked vessels were saved as their ships broke apart in the surf. Newspaper accounts concluded that at least thirty lives were lost in these shipwrecks, but the real number of deaths was probably much higher. A newspaper report from Norfolk, Virginia, following the August hurricane described the aftermath: "The stretch of beach between Kinnakeet to Hatteras, a distance of about eighteen miles, bears evidence of the fury of the gale in the shape of spars, masts, and general wreckage of five schooners which were washed ashore and then broken up by the fierce waves, while now and again a body washes ashore to lend added solemnity to the scene."

Of all of this hurricane's wrecks and rescues, one of the most dramatic was that of the barkentine *Priscilla.* This 643-ton American cargo vessel was commanded by Captain Benjamin E. Springsteen and was bound from its home port of Baltimore to Rio de Janeiro, Brazil. When the *Priscilla* left port on August 12, its captain was unaware of the fateful hurricane that would soon meet his ship head-on.

Wreck of the Priscilla *in August 1899. (Photo courtesy of the Outer Banks History Center)*

On the morning of Wednesday, the sixteenth, the wind began to blow, requiring that the ship's light sails be taken in. As the day advanced the winds continued to increase, and orders were given to take in all but the *Priscilla*'s mainsail. But by late afternoon, the driving wind had blown away or destroyed all of the vessel's riggings, and Captain Springsteen was now adrift under bare poles on a rapid southwest course.

Early on the morning of the seventeenth, after a stressful night of rolling seas and hurricane winds, soundings were made to test the water's depth. With each passing hour, the water became more shallow, and the captain knew that the storm was driving his ship ashore. Through the torrents of rain and wind, the order was passed to the crew to prepare to save themselves as the *Priscilla* was about to wreck.

After tossing about for the entire day, the ship finally struck bottom at about 9:00 P.M. on the seventeenth. For the next hour, the *Priscilla* was bashed against the shallow shoals as huge breakers crashed over its hull. Within moments, Captain Springsteen's wife, his son, and two crew members were swept overboard and drowned. Shortly afterward, and with a loud crash, the ship's hull broke apart, and the remaining horrified sailors held tightly to their wreck. Five more terrorizing hours would pass before the captain and his surviving crew would approach the beach.

Even though the hurricane's winds and tide were ferocious, Surfman Rasmus Midgett of the Gull Shoal Life-Saving Station set out on his routine beach patrol at 3:00 A.M. on the eighteenth. The ocean was sweeping completely across the narrow island, at times reaching the saddle girths of his horse.

But Midgett knew that disaster was at hand by the scattered debris that was washed about by the surf. Barrels, crates, buckets, and timbers provided clear signs that a wrecked ship was nearby. Although the night was dark and the storm was intense, this courageous surfman knew that lives were in jeopardy.

Finally, after an hour and a half of treacherous patrol, Midgett stopped on the dark beach at the sound of voices — the distressed cries of the shipwrecked men. Realizing that too much time would be lost if he returned to the station for help, he decided to attempt the rescue alone. One by one, he coaxed the *Priscilla*'s crew off the wreck and into the water, where he helped them to shore through the pounding breakers. Seven men were saved in this manner, and they gathered on the beach, exhausted.

Three of the crew remained on the wreck, however, too bruised and battered to move. Midgett swam out to save them and physically carried them to shore, one at a time. The courageous surfman brought the men to a high dune, where he left them to wait. His coat was offered to Captain Springsteen, who had received a serious wound to the chest. All of the men were bruised and bleeding, and some had their clothes stripped away by the relentless surf.

Midgett quickly returned to his station for help, and several men were dispatched to retrieve the survivors. In all, he had saved ten lives while risking his own in the treacherous waters of the San Ciriaco hurricane. For his efforts, he was later awarded a gold lifesaving medal of honor by the United States secretary of the treasury.

OCTOBER 30–31, 1899

The hurricane that battered the Cape Fear coast on Halloween Day 1899 was the second severe cyclone of the season to strike the Tar Heel state. Residents along the North Carolina coast had come to accept this kind of misfortune; it was the fourth time in less than twenty years that two major hurricanes had hit the state within a single year. Some even speculated that these violent storms were God's punishment to the citizens for allowing dancing on Sundays in local clubs.

The Halloween storm came ashore far to the south of where the San Ciriaco hurricane made landfall in August. And, like Hazel (1954), the Halloween hurricane struck the Brunswick beaches and cut a path through eastern North Carolina into Virginia. Even though the two storms of 1899 crossed different sections of the coast, their widespread effects brought great damage to some of the same locations. With the October hurricane, however, the destruction was most intense in the vicinity of Southport, Wilmington, and Wrightsville Beach.

In the late evening of October 30, the increasing wind and advancing tide offered late warning to the residents of Wrightsville and Carolina Beaches. Be-

cause the beach season had ended, most of the cottages in these resort communities stood empty, except for the caretakers and laborers who serviced them. As the storm rolled through in the early hours of the thirty-first, the greatest damage occurred before daylight. It was much later in the day before the hurricane passed and the people of Wilmington went out to survey the damage. A large crowd boarded the Seacoast train for Wrightsville to witness for themselves the severity of the storm.

When the train rounded the last curve before Wrightsville Station, those on board were not fully prepared for the vast destruction before them. The train screeched to a stop. One reporter from the *Wilmington Messenger* wrote:

> The massive railroad trestle was warped and twisted, and for a few hundred [feet] extending from the station towards the Hammocks the rails and ties were torn from the piles, and presented a tangled wreck piled down in the waters of the sound. The railroad tracks, approaching the station as far toward the city as the Pritchard cottage, was warped and torn and the large platform surrounding the depot was piled high with seaweed and other drift. To the right and left, stretching around the shore of the sound, as far as the eye could reach, where but yesterday, as it were, the famous shell road wound in beautiful curves, was a mass of deep tangled debris of every conceivable kind, the wreckage of cottages from the beach and of boats and bath houses along the shore of the sound.

During the peak of the storm, in many locations the ocean waves broke over the island into Banks Channel, carrying cottages with them. More than twenty cottages were either washed into the sound or completely wrecked by the

The Carolina Yacht Club was heavily damaged in the October hurricane of 1899. The structure was torn down after the storm and then rebuilt using much of the same lumber. (Photo courtesy of the Lower Cape Fear Historical Society)

Oceanfront destruction at Wrightsville Beach, October 1899. (Photo courtesy of the Lower Cape Fear Historical Society)

stormy surf. On the beach the railroad track south of Station Three was washed into Banks Channel, "the cross ties sticking up like a picket fence." Numerous shops, hotels, clubs, and docks were destroyed at Wrightsville. The public pavilion, the Ocean View Hotel, and the old Hewlett barroom were among the losses. The Carolina Yacht Club and the Atlantic Yacht Club were also severely damaged, both having been washed from their foundations and carried by the tides.

Several daring escapes were reported from Wrightsville during the height of the storm. The janitor in charge of the Carolina Club House abandoned the

Wreckage rests on the remains of the railroad tracks at Wrightsville following the 1899 hurricane. (Photo courtesy of the Lower Cape Fear Historical Society)

beach about midnight, before the hurricane reached full intensity. He jumped into a small skiff and sailed to his home in Hewlett's Creek, later describing his ordeal as "terrific in the extreme." Henry Brewington, a watchman for the Ocean View Hotel and several cottages, was forced onto the roof of the Russell cottage by the rising storm surge. When the dwelling collapsed, he swam through the breakers to the Atlantic Club. Realizing that this structure was also doomed, Brewington somehow made his way through the tide and back to the trestle at Wrightsville Station. J. T. Dooley, a railroad employee who lived on the Hammocks, narrowly escaped the rising waters with his wife and three children. At 3:30 A.M. he crossed the railway bridge with his family, and at 4:00 A.M. the trestle was destroyed by the pounding surf.

The destruction at Carolina Beach was equal to that of Wrightsville. The railroad tracks were damaged, numerous cottages were destroyed or "missing," and reports indicated that waves rolled through the town. Along the Wilmington waterfront, the Cape Fear River flooded the wharves from one end of the city to the other. Here, the water did not reach its high mark until 9:00 A.M., and then it fell rapidly. The flooding damaged stored goods all along the waterfront. As the waters rose inside the Atlantic and Yadkin Warehouse, a large quantity of lime became wet, causing fires to break out. The local fire department was called in to extinguish the flames. Somehow, a large flat of bricks was carried into the river by the storm, and "no trace of it has been found." It was also reported that "driftwood of every description, and seaweed were brought up the river in thick processions covering the entire surface of the river in some locations."

On the afternoon of October 31, a large trunk was found drifting in the river near the Wilmington docks. Inside were clothes, coins, and papers belonging to J. W. Brock. Because Brock was known to have been fishing at Zeke's Island prior to the storm, there was speculation that his fishing party was washed away when the island was "completely covered" by the hurricane's surge. Brock was never heard from.

The damage at Southport caused one resident to report: "The storm here was the worst ever known and great damage was done." The storm tide rose five feet above normal high water, and many houses along the waterfront were badly damaged. The Norwegian bark *Johannah* was being "disinfected" at the Cape Fear quarantine station when the storm approached. The entire quarantine crew took refuge on the ship, which broke anchor and washed aground up the river. The Southport waterfront was littered with the wreckage of the hurricane, including numerous small boats, the remains of the city wharves, and the passenger steamer *Southport*. According to the *Messenger*, "Large droves of cattle drifted across the river, dead and alive. They were run off by high waters all over Bald Head Island, which never was known to be covered before."

The hurricane struck north of the Cape Fear region as well. The New River Inn and over a dozen cottages were swept away in Onslow County. One report claimed that all the oysters in New River were covered by sand deposited by the tides, and none could be harvested for years. Also, Old Stump Inlet, said to have been closed for more than a generation, was reopened by this storm and was reported to have had twelve feet of water on its bar.

Farther up the coast in Morehead City and Beaufort, the effects of the Halloween hurricane were as unwelcomed as the tragedies that had struck with the San Ciriaco storm. High water again flooded the low-lying reaches of these coastal towns. In New Bern, the storm brought that city "the worst experience in her history." The water was two feet higher than during the August hurricane, and once again fires caused by wet lime broke out on the docks. Few reports were offered from other coastal villages, such as Portsmouth and Ocracoke, but the Halloween hurricane was likely to have been felt along much of the Carolina coast.

HURRICANES OF THE NEW CENTURY,
1900–1950

NOVEMBER 13, 1904

The first severe hurricane of the twentieth century to move across the North Carolina coast came late in 1904, on November 13. This category-three storm passed near Hatteras during the morning and brought high tides and heavy rains to the entire coast. Two schooners were wrecked near Cape Fear, and extensive damage was reported at Fort Caswell. At New Inlet, the storm surge swept away the Life-Saving Station, and four crewmen drowned. Four more lives were lost in the wreck of the *Missouri*, a schooner that went down near Washington, North Carolina. Several lives were lost when a fishing lodge on Hatteras Island was swept away by the tides, and eight more people drowned when a yacht sank in Pamlico Sound. As the hurricane moved past the coast, a large cold-air mass was drawn into the cyclone's circulation, and an early snow fell across much of the state.

SEPTEMBER 17, 1906

On September 17, 1906, a severe hurricane made landfall near Myrtle Beach and caused considerable damage along the southern coast. Although the winds in Wilmington reached only 50 mph, the tides were high and breakers were reported in the streets of Wrightsville Beach. Just as in the Halloween hurricane of 1899, the trolley car trestle to Wrightsville was damaged. Several cottages and a hotel were washed away, and more than two hundred people were rescued by boat. One unusual account from this storm came from Masonboro, where Walter Parsley reported finding large bowling balls in his front yard after the storm. Although the balls were made from a dense wood called *lignum vitae*, they had somehow washed across the sound from Wrightsville.

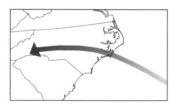

SEPTEMBER 3, 1913

On the morning of September 3, 1913, a short-lived but severe hurricane crossed over Core Banks into Pamlico Sound. It decreased in intensity as it continued on a west-northwest path, passing south of Raleigh in the afternoon. Although its fury could not be compared with the "great" hurricanes, this storm brought surprising amounts of floodwaters and destruction to several down-east locations. A maximum wind velocity of 74 mph was reported at Cape Hatteras, but this storm's greatest impact occurred when the wind-driven waters of Pamlico Sound were pushed inland.

The most severe flooding was reported in New Bern and Washington. The streets of New Bern were inundated by the overflowing Neuse River, which reached a new record level for that location: nine feet above normal high water. A large railroad bridge that spanned the Neuse was washed away, and

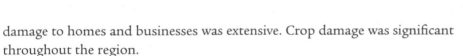

The Washington, N.C., waterfront after the destructive flood of 1913. (Photo courtesy of the George H. and Laura E. Brown Library)

New Bern residents look over the water and debris filling Johnson Street after the September hurricane of 1913. (Photo courtesy of the North Carolina Maritime Museum)

damage to homes and businesses was extensive. Crop damage was significant throughout the region.

In Washington, the water rose ten feet above normal, and the flooding extended all the way to Third Street, which is the fourth street from the river. There was so much water on Main Street that "speed boats were coming and going" most of the following day. The Norfolk and Southern trestle, the Washington and Vandemere trestle, and the old county bridge were all washed away and had to be rebuilt the year following the storm.

As in past hurricanes, several ships were reported as wrecked or lost. These included the *Dewey*, which sank at Cape Lookout, the schooners *Manteo* and *Grace G. Bennett*, which became stranded near Portsmouth, and the Boston schooner *George W. Wells*, which wrecked about five hundred yards offshore at Ocracoke. At 317 feet and almost three thousand tons, the *Wells* was hailed as the largest sailing vessel of its kind in the world. In a terrifying fourteen-hour ordeal, the passengers and crew of the *Wells* were saved by the courageous men of the Hatteras Inlet Life-Saving Station.

The hurricane of 1913 may have been only a category-one storm, but its westerly movement through Pamlico Sound and across the state brought great destruction to a broad area. Crops were damaged and structures were destroyed throughout eastern North Carolina. At Goldsboro, the storm was "the worst in history," and similar reports came from Tarboro, Wilson, Farmville, and Durham. In all, five lives were lost and property damage was estimated at $3 million.

JULY 14–16, 1916

Although the midsummer hurricane of 1916 actually moved ashore on the South Carolina coast, its northwest movement carried it inland through the North Carolina mountains on July 15. There the storm dumped record amounts of rain across the southern Appalachians. The heaviest rains were recorded at Altapass, where 22.22 inches fell during the twenty-four-hour period ending at 2:00 P.M. on the sixteenth. This downpour established a new twenty-four-hour rainfall record for the entire United States.

Winds from the storm were less than hurricane strength in North Carolina, but the deluge of rain brought deadly landslides to the mountain countryside. Bridges and highways were washed away, and crops suffered great damage. Railroads were undermined in numerous locations, shutting down the lifelines of rail transport. Several persons were killed in the mudslides that followed the record rains. No significant damage occurred along the North Carolina coast.

SEPTEMBER 18–19, 1928

The Great Hurricane of 1928 brought epic destruction to Puerto Rico and Florida before it tracked northward through the Carolinas. On September 13, the hurricane moved over Puerto Rico, where winds were clocked at 160 mph and 300 lives were lost. As the great storm moved over eastern Florida on the seventeenth, the wind-driven waters of Lake Okeechobee overflowed into populated areas, causing a massive flood. In Florida, 1,836 were killed, mostly due to drowning.

By the time the hurricane crossed into the sandhills of North Carolina on the night of September 18, its destructive winds had diminished, but tremendous rains fell across the Tar Heel state. The resulting floods were very severe and the highest on record for some upper portions of the Cape Fear River. At Fayetteville, where the bank-full stage is 35 feet, the river reached an unprecedented height of 64.7 feet. At Elizabethtown, the river rose to 41.3 feet. Flooding at Lumberton was reported as "the worst in history," and thousands of acres of crop lands were underwater. Many highways were closed because of bridge washouts and deep-standing water.

OCTOBER 1–2, 1929

Another major hurricane struck Florida and tracked northward through the Carolinas in 1929. Like the storm of the previous year, the 1929 hurricane lost much of its wind energy before it reached the south-central region of North Carolina. But the storm's massive rain clouds emptied as it crossed the state,

Oceanfront cottages at Wrightsville Beach were left exposed following the October hurricane of 1929. (Photo courtesy of the New Hanover County Public Library)

again causing severe river flooding and crop damage. Record rainfall amounts were reported in many locations. Flooding on the Cape Fear was very near the record level of the previous year, and at one station near Fayetteville the river rose forty-one feet in one twenty-four-hour period. For the second straight year, many portions of North Carolina "foundered in flood."

AUGUST 22–23, 1933

By 1932, the Great Depression had a firm grip on the nation and had cast its long shadow across North Carolina. Times were hard from the Appalachians to the coastal plains. In November of that year, Franklin D. Roosevelt was elected president with the promise of a New Deal for economic recovery. By the summer of 1933, there seemed to be reasonable hope among the people of the nation. But for many coastal North Carolinians, the months of August and September would bring a far greater despair than the recent economic turmoil.

Once again North Carolina was struck by two hurricanes within a single season. On August 22 and again on September 15, the coastal region was awash with storms. The August hurricane, ranked as a category two in intensity, passed over the northern Outer Banks, and in September a deadly category-three hurricane spun through Pamlico Sound and up the coast.

The first of the two storms passed east of Ocracoke just after midnight on the morning of August 23. By midmorning, the hurricane was on a curving

path back toward the Atlantic by way of Norfolk, Virginia. At Cape Hatteras, the maximum winds were only 64 mph. But high tides and severe beach erosion were reported all along the banks. Crops were damaged as far inland as Granville County, and Norfolk, Virginia, was flooded by "several feet" of water. In the northeastern counties, the damage was estimated at $250,000.

SEPTEMBER 15–16, 1933

Less than one month after the August storm, another hurricane moved toward the coast. By the morning of September 15, it had reached a position 250 miles south of Cape Hatteras, and all indications were that it would cross the North Carolina coastline late that night or early the next morning. Storm warnings were issued, and the few coastal cities that had hurricane plans put them into effect. The American Red Cross, better organized after its recent experience with tragic hurricanes in Florida, urged the Carolina chapters to prepare for a potential disaster.

When the hurricane approached the mainland, it swerved to the north. As it pushed through Pamlico Sound, intense northeast winds forced tremendous quantities of water to surge to the southwest, flooding the river basins of the Neuse and the Pamlico. An unusual phenomenon occurred along the northern banks of the Albemarle Sound when the water was "blown away" to the lowest level ever recorded for that region.

The tremendous storm tide that swept through several down-east communities claimed twenty-one lives and left extensive destruction. The wind-driven water remained high on the land until the storm moved up the coast.

Then, like a cork removed from a bathtub, the water rushed back toward the sea, overwashing Core Banks from west to east and opening Drum Inlet in the process. Winds were recorded at 92 mph at Cape Hatteras, just before the anemometer was destroyed. In Beaufort and New Bern, winds were estimated at up to 125 mph. In some areas, wind-related damage was as severe as the flooding. Countless large trees were downed throughout the east, including the city of New Bern. In an article from the *New York Times*, a reporter wrote: "New Bern has long been known as the 'Athens of North Carolina' because of its many large and beautiful trees. Now hundreds of these trees are either lying in the streets or leaning grotesquely against the battered houses. Many of the splendid trees of East Front, Broad, Pollock, Johnson, and Craven streets were blown down."

The flooding in New Bern was the highest ever known and was said to have been about two feet greater than in the storm tide of September 1913. The water reached a height of three to four feet in some streets, and rowboats and skiffs were used to evacuate people from buildings that were completely surrounded by water. The tide rose a foot above the tallest piling on the Coast Guard dock, and the dock was wrecked and washed away. The cutter *Pamlico* was unharmed, however, even though it was moored to the dock at the time.

The Neuse River bridge that linked New Bern and Bridgeton on U.S. 17 was washed out at about 1:30 A.M. on September 16. A three-quarter-mile-long section was taken out by the surging waters of the Neuse, and pieces of the bridge were scattered along the shore for miles downriver. Two stalled automobiles were believed to have been stranded on this section, but their occupants were able to escape the bridge before it collapsed. Damage was also reported to the Norfolk and Southern railroad trestle and the Trent River bridge. Several boxcars were dumped into the Neuse when the Atlantic Coastline pier caved in.

One unusual story from the "storm of '33" came from the New Bern area. The roof of Mrs. Sam Smallwood's boathouse was blown off by the fierce winds. It landed, right side up, on a seawall a quarter mile downriver, with Mrs. Smallwood's boat, which had been suspended from the rafters, still intact. The boat was retrieved the morning of the storm and was used to take Mrs. Smallwood to higher ground.

The damage in New Bern alone was more than $1 million. Many homes and businesses were severely damaged by both high water and winds. Several lumber factories were severely damaged. The rising tides of salt water reached the region's farmlands, and damage was heavy to unharvested corn, cotton, and sweet potatoes. In many locations, tons of tobacco stored in barns were destroyed when they became soaked by heavy rains and rising floodwaters. But by far, the greatest loss and suffering came to those who lived near the water and made their living from the sea. Carteret County was hit hard by the storm, and remote communities down east were devastated.

The *Beaufort News* reported the following events the week following the storm:

The oldest citizens here in Beaufort have told the News that it was the most devastating storm that they have seen in the past four score years. It was not merely a bad wind that reached gale force for just a few minutes; the disastrous hurricane swept Carteret for more than twelve hours without ceasing for even a few minutes. From early Friday morning rain began falling and this continued unremittingly until about day break Saturday morning.

This terrific tropical hurricane which swept up the Atlantic coast Friday seemed to have hit Carteret near Beaufort Inlet, striking Beaufort and Morehead City first, then continued with its destructive force on to Merrimon, South River, Lukens, Roe and Lola, with all other communities in eastern Carteret getting their shares of the devastating tempest. . . . Within Carteret County alone there was a property loss of at least a million dollars, eight people were drowned and scores left homeless, hundreds without food and more with barely enough clothing to cover their bodies. Thousands of domestic and wild animals perished in the water and if they are not removed and buried decomposition will result in stench and disease. In the villages where homes and other buildings were wholly or partially demolished, men, women and children by the score stuck nails in their feet and have cuts and bruises and sprains across their bodies. Only a very small percentage have received medical attention and been inoculated with tetanus antitoxin. Sanitary conditions in the stricken area are terrible, and epidemics will in all likelihood ensue if the people do not cooperate whole-heartedly with the sanitary engineers of the State Department of Health.

Captain Jim Hamilton and his three sons, Nelson, Charlie, and Ralph, all drowned in Long Bay when their twenty-foot skiff capsized in the storm. Like countless other fishermen before them, they had left their home in Sea Level with no knowledge of the impending weather. Their expedition quickly turned tragic as their small boat was no match for the furious seas.

In the down-east community of Merrimon, the tide was estimated at "fifteen or sixteen feet." Only four out of thirty houses remained after the tides overwashed the area. The Carraway family endured a horrible ordeal when their house collapsed during the storm. The entire family huddled together as a blast of wind tore down the structure, pinning them in the wreckage and the rising tides. Mr. Carraway escaped with the help of his son George, but Freda Carraway remained trapped under the debris. Those who escaped were forced to flee to higher ground when the tides continued to rise, but Freda remained trapped under the house, where she apparently drowned.

At nearby Cedar Island, about eight families endured the hurricane, and almost all of their homes were washed off their foundations or severely damaged. In his book *Sailin' with Grandpa*, author Sonny Williamson offers a detailed account of the scene as recorded by Captain John Day and his wife, Adelaide. Day had come to Cedar Island by boat to check on his relatives in the remote community. Virtually every structure suffered damage in the storm, and the bewildered residents struggled to simply survive in the first days following their ordeal.

After the hurricane passed, the local director of the Red Cross, Frank Hyde, launched a relief mission for the down-east villages. No communication was possible with the isolated hurricane victims, and this voyage would provide the first news of the conditions in these areas. Thirty "orders" of food and supplies were prepared for the relief effort, which was assisted by the Coast Guard. Hyde left Fort Macon for Core Sound early Monday morning, two days after the hurricane's arrival.

The mission reached Lola, on the southern end of Cedar Island, by late morning. James Whitehurst, reporter for the *Beaufort News*, was traveling with Dr. Hyde and filed the following report:

Upon arriving on shore we were conducted through a throng of half-clothed bewildered people who looked upon us with overjoyed eyes. One young woman with a baby—it appeared to be her first—cried with joy. Every person seemed to have stuck nails in their feet or had cuts and bruises about their bodies. The last food in Lola had been consumed for breakfast, and this had been far from sufficient.

The homes had been washed from their foundations, windows had been blown out, roofs and roofing wrenched from the tops of the structures. Wreckage was strewn from one end of the island to the other. Few of the people had shoes on, and virtually every one had on all the clothing they had been able to salvage.

When the Coast Guard boat carrying Dr. Hyde moved to the northern end of the island, its passengers found even greater destruction in the village of Roe. Eight or ten homes were described as totally destroyed, and only one was "fit for winter." Most of the homes had floated haphazardly with the tides, and many suffered structural damage. Thick mattes of mud, grass, and debris filled several houses and littered the branches of nearby trees.

At South River, similar floods struck late Friday night. The Louis Cannon family narrowly escaped drowning when their home collapsed in the storm. They had gathered in their attic when the waters rushed into the first floor. They eventually escaped by clinging to the rooftop of the broken house until they became caught in the top of a grove of trees. There they rode out the storm until the waters receded.

The home of William Cannon became a refuge for other South River residents who were forced to flee their homes. The Cannon residence was on higher ground than many other houses, and frightened neighbors and relatives made their way there when the floods moved in. Ultimately, more than fifty people were sheltered in this one house. Many stayed through Saturday, until they were able to return to what was left of their homes.

One of the great tragedies of the '33 storm struck the family of Elijah Dixon, who were staying in a two-story home near Back Creek when the hurricane hit. Dixon, his wife Ellen, their eight-year-old daughter Hazel, three-year-old son James, and nine-month-old daughter Elva Marie were all plunged into the raging waters when the house washed into Back Creek. The family tried desperately to cling to the broken fragments of the rooftop. With his young son around his neck, Dixon jumped into the dark waters to rescue his wife, who was still clinging onto baby Elva. In the darkness and confusion, the infant slipped from her arms and drowned. As the weary group again gathered on the roof, they realized that young Hazel was also missing. Reeling from this double tragedy, Dixon still managed to grasp a large branch when the rooftop was swept into a grove of trees. There the battered family remained until later Saturday afternoon, when they were rescued.

The tragedies of the 1933 hurricane were spread throughout numerous down-east communities. At Oriental, Vandemere, Bayboro, and Arapahoe, local newspapers reported that "hardly a building was left intact." From Ocracoke, there was a report that "four feet of water had covered the island."

As the hurricane's storm surge pushed over Ocracoke, residents scrambled for high ground in any way they could. Some had prepared their homes by removing floorboards to prevent the tides from washing their houses "off the blocks." Many rode out the hurricane in their attics, and some were forced onto their rooftops.

At the Green Island Hunt Club, on the eastern end of Ocracoke, nine people took refuge as the storm approached. The two-story clubhouse was rocked by pounding surf, and by midnight, "the ocean was breaking in the kitchen." As the first floor filled with water, the occupants gathered on the second floor, which also began to flood. Like so many other storm victims throughout the region, the group was forced to crawl through a high window onto the roof, where they rode out the storm. The structure pitched and rolled like a boat, and the constant up and down motion washed away the foundation, digging a deep hole in the sand. The frightened men and women held desperately to the rooftop through the gusting winds and strong surf until, at daybreak, the storm passed.

As the tides receded, the survivors climbed down off the clubhouse. To their amazement, the wave action had dug the house so far down that the second-floor windows were level with the ground. The fishermen were also surprised to see their boats cast upon a nearby beach. During the storm, these small craft were blown out into the sound, but when the hurricane passed and the winds shifted, they were blown right back, suffering little damage.

The September hurricane of 1933 left many scars on the North Carolina coast. In all, twenty-one were dead and damage estimates topped $3 million. The Red Cross estimated the area of greatest suffering to be a nine-county region with a population of 120,000. Their survey indicated that 1,166 buildings had been totally destroyed and 7,244 severely damaged. The Red Cross gave aid to 1,281 families, many of whom received help with the rebuilding of their homes. This effort helped temper the anguish from one of the most tragic storms in North Carolina history.

SEPTEMBER 18, 1936

The residents of North Carolina's Outer Banks suffered through another severe hurricane on September 18, 1936. Like so many storms, this hurricane swept along the banks with its eye just offshore, curving back out to sea before its intense eastern half could reach land. Even though the eye remained over the ocean, the hurricane was quite severe, especially along the northern coast. Winds of 90 mph were reported at Manteo, and Hatteras station recorded an average five-minute wind speed of 80 mph.

The strong northeast winds brought significant beach erosion to the coast. About thirty-five feet of beach were lost at Nags Head. Many roads and highways were covered with water and sand, and bridges were undercut by the

Many of Southport's riverfront docks and shrimp houses were destroyed in the August hurricane of 1944. (Photo courtesy of Leila H. Pigott)

tides. The highway from Currituck to Norfolk was washed out, hampering relief efforts after the storm.

On Hatteras Island, residents once again had to reckon with the "hoigh toide on the sound soide." Rising water washed numerous homes off their foundations. Some families opened their doors to the storm, allowing floodwaters to wash through their houses, thus preventing the structures from being swept away. After the tides receded, brooms were used to sweep out the heavy silt that had collected on the floors and baseboards.

The Hatteras residents were threatened by disease after the '36 hurricane, and drinking water was scarce. Large cisterns used to collect rainwater were contaminated after the storm, and each one had to be disinfected with bleach. Outdoor toilets were common, and most had overflowed in the flood, contaminating the drinking water and presenting the threat of disease. Typhoid vaccine was administered by a local doctor to most residents. Until the cisterns could be cleaned and more rains could come, many of the people living on the island had no water to drink. Fortunately, the American Red Cross brought in drinking water for a period of weeks after the storm.

The 1936 hurricane was severe from Cape Hatteras through southern Virginia. But fortunately, this storm was not as destructive (it ranks as a category two) as it might have been had it continued on its northwest track. Nevertheless, it still caused $1.6 million in damages in North Carolina and Virginia and claimed two lives.

AUGUST 1, 1944

The 1944 hurricane season was active along the East Coast, and North Carolina felt the effects of two hurricanes that year. The first was a storm of small diameter that moved northward past the Bahamas on July 30 and made landfall near Southport at about 8:00 P.M. on August 1. This category-one hurricane brought winds of 80 mph to Oak Island, and gusts in Wilmington reached 72 mph. The greatest reports of damage came from Carolina Beach, although the waterfront in Southport suffered severely.

A remarkable evacuation of island residents and vacationers was completed at Wrightsville and Carolina Beaches. It was estimated that ten thousand people were removed from these resorts, and many were taken out in the desperate hours of the storm's arrival. One hundred army trucks were brought in from nearby Camp Davis to assist with the transport as army, Coast Guard, and police officers went door to door calling for residents to leave. Many vacationers attempted to flee in their cars and were trapped in rising waters that stalled their vehicles. The Wrightsville Causeway and Carolina Beach road were flooded, and stalled cars blocked the transport of evacuees. Soldiers and police worked quickly to push aside the cars and successfully complete the evacuation.

During the peak of the storm, a two-and-a-half-ton amphibious vehicle was called in from Camp Davis to rescue seven stranded soldiers from Hogshead Island in the Cape Fear River. The "duck" was driven into the river through towering waves and reached the island within a few minutes. The soldiers, stationed at nearby Fort Fisher, would likely have drowned had they not been rescued, according to their commanding officer.

Although it is possible the winds never reached 100 mph, trees and power lines were downed throughout the region. Large plate-glass windows were blown out in downtown Wilmington, and some of the city's most majestic oaks were toppled. No deaths were reported, but there were several injuries, including two young boys who were burned by downed power lines. A local paper boldly reported that this was "the greatest storm to strike here in the past 200 years."

The most extensive damage occurred at Carolina Beach. Thirty-foot waves reportedly pounded the beachfront and totally destroyed the town's famed boardwalk. According to the *Wilmington Morning Star*, oceanfront homes were washed from their foundations, which "left them at crazy tilts like the hats of drunken sailors weaving down the street." At Wrightsville, the greatest damage occurred to the new sewage project, which was under construction and was left completely covered by sand. Police officers who remained at Wrightsville through the storm reported that "at one time, the water measured 18 feet by the City Hall." The total damage from the storm exceeded $2 million.

SEPTEMBER 14, 1944

The second hurricane of 1944 to strike North Carolina brought death and destruction to nine hundred miles of coastline from Cape Hatteras to Newfoundland. The Great Atlantic Hurricane of 1944 approached the Outer Banks on the morning of September 14 and passed just east of Cape Hatteras on a track toward New England. The eye of the hurricane remained offshore, and North Carolina's coast was spared a direct hit. But as the great storm moved northward, its forward speed increased, and it slammed into Long Island, New York, at about 10:00 P.M. the same day. The damage was heavy in New York, Connecticut, Rhode Island, Massachusetts, and Maine.

The pattern and destruction of the '44 storm was very similar to that of the Great New England Hurricane of 1938. The '38 storm still ranks as one of our nation's worst natural disasters. The American Red Cross reported that it claimed 494 lives (more than another 100 were lost at sea), destroyed over four thousand homes and buildings, and brought $350 million in damages to the northeastern states. Although the 1938 hurricane brought no serious effects to North Carolina, it did blow by Hatteras and then accelerate to the north, reaching an incredible forward speed of 56 mph. This meteoric movement intensified the storm's effects on the New England coastline, and the hurricane's surge drowned scores of coastal residents. During the 1938 hurricane, one of the highest wind speeds in U.S. history was recorded in Milton, Massachusetts—a gust measuring 186 mph.

Most of the homes and businesses destroyed by the Great New England Hurricane of 1938 had been rebuilt by the summer of 1944. But the Great Atlantic Storm of '44 again flooded the northeast, and damage from this hurricane was heavy. In North Carolina, the '44 storm was called the "worst ever" by some on the Outer Banks, who again watched as the ocean and sound met, surrounding their homes with water. On Ocracoke, the tide rose two to four feet in many houses. As they had done so many times in the past, residents there opened their doors to the rising waters to keep their homes in place. Fish were reportedly trapped under furniture and left behind when the storm passed and the waters receded.

The hurricane moved northward up the banks at about 30 mph on the morning of September 14. At Cape Hatteras, the weather station recorded a barometric pressure reading of 27.97 inches, the lowest to date for that location. Even though its center remained at sea, the storm's massive eye wobbled up Hatteras Island and was discernible to the island's residents. It was reported that the eye was so large that many had time to leave their homes to check on their boats, visit friends, and survey the damages in the area. Some were caught off guard when the hurricane's eye passed by and winds exceeding 100 mph "hauled around," blowing from the northwest.

During the storm's approach, strong southeast winds had filled the sounds with ocean water, backing up all the rivers, creeks, and marshes on the mainland side. Some on Hatteras Island reported that the winds had blown the waters so far west that the sound was left dry for nearly a mile. But as the eye passed and the winds turned around, the waters of Pamlico and Albemarle Sounds rushed back toward the banks, flooding villages from Hatteras to Nags Head. Time and again this phenomenon of wind and water recurs with passing hurricanes, each time bringing new potential to reshape the Outer Banks.

The most serious flooding was in Avon. During the 1930s, the Civilian Conservation Corps had installed miles of fencing on the beach, which collected sand to form a protective barrier. The "sand walls" practically surrounded the town. When the '44 hurricane's west winds pushed the sound into the village, the sand walls became a dike that prevented the waters from escaping. The massive surge could not continue on across the bank to the ocean, and Avon became a deep pool. Residents reported seeing cars and trucks completely covered by the tides. Houses drifted about, sometimes crashing into each other. One young girl, huddled in desperation with her family in a second-story hallway, actually became seasick as her home was sloshed about by the tide and wind. Avon was hit hard, as 96 of the town's 115 houses were severely damaged or washed off their foundations.

Along the northern banks, the pounding northeast winds and waves scoured out the skeletons of forgotten shipwrecks, some of which had been buried for generations. There was heavy damage in Elizabeth City and in the Nags Head area, some from flooding but much from the storm's high winds. At Cape Hatteras, winds were estimated at 110 mph (after the anemometer failed), but at Cape Henry, Virginia, the hurricane's top speed was recorded at 134 mph. The highest gust at that location was estimated at 150 mph. Maximum wind velocities equaled or exceeded all previous records at Hatteras, Cape Henry, Atlantic City, New Jersey, New York City, and Block Island, Rhode Island.

Although only one North Carolinian was killed in the Great Atlantic Hurricane of 1944, the storm was deadly overall. In the northeast, 45 more lives were lost, including 26 in Massachusetts. But most tragically, 344 more people died at sea as five ships sank during the hurricane. Two of those were lost off of the North Carolina coast. The Coast Guard cutters *Jackson* and *Bedloe* both capsized and sank while guarding a liberty ship that had been torpedoed near Cape Hatteras. German U-boats were active at that time in attacking allied vessels that moved along the Outer Banks. The '44 hurricane provided only a brief interruption in their acts of war.

HURRICANE ALLEY,
1950–1960

7

During the fifties, while most of America was laughing at Milton Berle, rockin' with Elvis, or worrying about Khrushchev and the bomb, the people of eastern North Carolina were grappling with a different kind of surprise attack —hurricanes. Throughout history these violent storms had visited the Carolina coastline with random frequency. Sometimes they had struck twice or even three times within a single year. Most often, however, several years had passed between major storms, and coastal residents had been able to pick up the pieces of their lives and rebuild before the next hurricane. But for a period in the mid-1950s, a flurry of hurricane activity dispelled all statistical expectations. Seven hurricanes blasted the Tar Heel state in roughly two years, leaving death and great destruction in their wake. Included in this group was the most infamous of all North Carolina hurricanes, the "lady" known as Hazel. The great destruction brought on by this group of storms earned a new nickname for the region: Hurricane Alley.

BARBARA (AUGUST 13, 1953)

In 1953 the United States Weather Bureau officially began providing women's names for hurricanes in the Atlantic. Hurricane Barbara, the second-named storm of that season, struck the coast of North Carolina between Morehead City and Ocracoke on August 13. After spinning northward along the Outer Banks, the storm turned to the northeast and moved out to sea near the Virginia line. Barbara was a category-one hurricane, and damages were not severe. The storm's highest winds were reported as gusts to 90 mph at Cape Hatteras and Nags Head. Several locations along the coast reported rainfall exceeding six inches, but winds and rains on the mainland were very light. Damage estimates exceeded $1 million, but most of this was attributed to crop damage. The only death associated with Barbara occurred at Wrightsville Beach, where a man was swept from a pier and drowned.

CAROL (AUGUST 30, 1954)

One year later, on August 30, 1954, hurricane Carol accelerated up the Atlantic coast, just missing Cape Hatteras. The hurricane's eye passed the cape at about 10:00 P.M. on a north-northeast track. Because the Outer Banks remained on the storm's weaker western side, damages were not severe. Beach erosion was significant, and some homes and fishing piers were damaged. About one thousand feet of highway were undermined on the banks. Although property damage at any given locality was light, damage spread along the coast totaled $250,000. Carol's highest winds were between 90 and 100 mph at Cape Hatteras. Gusts were recorded to 65 mph at Cherry Point and 55 mph at Wilmington. Tides ran from three to five feet above normal.

Homeowners and merchants whose properties faced the ocean suffered the greatest losses from the hurricanes of the 1950s. (Photo courtesy of the News and Observer Publishing Co./N.C. Division of Archives and History)

Residents of the coastal Carolinas faced the recurring ordeal of hurricanes for an extended period during the 1950s. (Photo courtesy of the News and Observer Publishing Co./N.C. Division of Archives and History)

North Carolina was spared the full brunt of hurricane Carol, but residents of Connecticut, Rhode Island, and Massachusetts were not as fortunate. Carol whipped across the Northeast at a torrid pace, with a forward speed of almost 40 mph. Winds at Block Island, Rhode Island, reached 100 mph, with gusts to 130. Tides at Providence were eight to ten feet above normal, and damage to waterfront structures was heavy. The storm surge and high winds were responsible for sixty deaths in the New England states. Property damages from Carol totaled $461 million, which made this hurricane the most expensive in U.S. history to that date.

EDNA (SEPTEMBER 10, 1954)

Less than two weeks after Carol had passed, residents along the North Carolina coast were told that another hurricane was headed their way. Hurricane Edna followed a track very similar to that of Carol, sweeping past the Bahama Islands and curving northward toward the Tar Heel coast. Fortunately, Edna also passed offshore, missing Cape Hatteras by some sixty miles.

Once again, damages were generally light but widespread along the Outer

A sense of bewilderment came over many who searched for the remains of their homes at Long Beach following hurricane Hazel in October 1954. (Photo courtesy of Lewis J. Hardee)

Banks. Television aerials, roofs, and fishing piers were damaged from Ocracoke to the Virginia line. Another section of highway was washed out on Hatteras Island. Beach erosion was severe in some areas due to large waves and strong currents brought on by northeast winds that gusted to 75 mph.

But like Carol, hurricane Edna missed the North Carolina coast and sped northward to strike the northeastern states. Edna made landfall near the eastern tip of Long Island, New York, on September 11. Winds gusted to 120 mph at Martha's Vineyard, Massachusetts, and a great storm tide again flooded the region. The death toll from this storm reached twenty-one and damages exceeded $40 million. Edna landed a knockout punch to numerous coastal communities that were still reeling from the devastation left by hurricane Carol. The 1954 hurricane season turned out to be one of the most destructive in New England history.

HAZEL (OCTOBER 15, 1954)

Carol and Edna had given the residents of coastal North Carolina a brief scare and a taste of nasty weather. But as it turned out, these two storms were only warning shots fired beyond the coastline. The next hurricane took dead aim on Brunswick County and would eventually be recognized as one of the greatest natural disasters ever to affect the state. Its name was Hazel.

Hazel began, like many other hurricanes, as a trough of low pressure over the warm waters of the tropical Atlantic Ocean. On October 5, 1954, the small storm was identified just east of Grenada and was observed on a west-northwest track that would take it through the Grenadines. Winds were clocked at 95 mph, and the small island of Carriacou was the first to suffer from Hazel's winds and tides.

At the Weather Bureau's Miami office, chief forecaster Grady Norton was working twelve-hour days plotting Hazel's course. Tragically, Norton suffered a stroke on the morning of October 9 and died later the same day. The Weather Bureau scrambled to pick up where Norton had left off, following the daily reports of the growing storm. Although he had been warned by his doctor to avoid the long hours, Norton had ignored his medical condition out of concern for the hurricane's potential for destruction.

Hazel continued to draw energy from the warm Caribbean waters and intensified as it curved northward toward Haiti. During the early morning hours of October 12, the well-organized storm slammed into the Haitian coast, raking over both the southern and northern peninsulas.

Haiti was devastated. Several small towns were "almost totally demolished," and larger cities such as Jeremie and Port de Paix suffered severe wind damage. Torrential rains fell over the island nation, flooding rivers, washing out roads and bridges, and filling homes and businesses with mud. Heavy rains caused a massive landslide that buried the mountain village of Berley, killing almost all of its 260 residents. Winds in excess of 125 mph were reported from several Haitian cities. Tides reached record levels on the southern peninsula, and surge flooding was extreme. It was estimated that as many as one thousand Haitians may have died in the storm.

As Hazel moved through the Windward Passage on the morning of October 13, observers noticed that its winds had diminished to a mere 40 mph. This low wind speed was the result of a distortion of the hurricane as it passed over the mountainous Haitian terrain. But Hazel rapidly regained its strength and form over the next few hours. After passing the Bahamas, it was again a major cyclone, and all warning messages issued for the United States coast indicated sustained winds in excess of 100 mph.

The eye of the hurricane passed about ninety-five miles east of Charleston, South Carolina, at 8:00 A.M. on October 15. About this same time, the outer fringes of the storm first touched the U.S. coastline near North Island, South Carolina. Finally, between 9:30 and 10:00 A.M., Hazel's ominous eye swept inland very near the North Carolina–South Carolina border. It was there on the coast that its most awesome forces were unleashed, although the destruction continued as the hurricane barreled across the state.

The storm surge that Hazel delivered to the southern beaches was the greatest in North Carolina's recorded history. The flood reached eighteen feet above mean low water at Calabash. Hazel's surge was made worse by a matter of pure coincidence—it had struck at the exact time of the highest lunar tide of the year—the full moon of October. Local hunters often refer to this as the

"marsh hen tide," a time when high waters tend to flush waterfowl out of the protective cover of the marsh grass. Hazel's storm tide may have been boosted several feet by the unfortunate timing of its approach.

The coastal region where Hazel made landfall was also battered by some of the most destructive winds in North Carolina's history. Estimates of 150-mph extremes were reported from several locations, including Holden Beach, Calabash, and Little River Inlet. Winds of 98 mph were measured in Wilmington and were estimated at 125 mph at Wrightsville Beach and 140 mph at Oak Island. As Hazel swept inland, its winds endured with freakish intensity. Grannis Airport in Fayetteville reported gusts of 110 mph, and estimates of 120 mph were made by observers in Goldsboro, Kinston, and Faison. At the Raleigh-Durham Airport, the wind-speed dial was watched closely during the storm, and gusts to 90 mph were recorded around 1:30 P.M. Most incredibly, wind gusts near 100 mph were reported from numerous locations in Virginia, Maryland, Pennsylvania, Delaware, New Jersey, and New York as Hazel curved a path through the Northeast on its way to Canada.

Hazel's violent winds hacked or toppled countless trees across eastern North Carolina. In the aftermath of the storm, some sections of highway were littered with "hundreds of trees per mile." Some were uprooted and tossed about, and others were snapped off ten to twenty feet above the ground. In the city of Raleigh, it was reported that an average of two or three trees per block fell. Many fell on cars, homes, and other structures, and power lines were left tangled and broken. Dozens of other cities and towns in the eastern half of the state faced similar losses.

Effects of the hurricane as it moved inland were remarkable. According to a report from the National Weather Service,

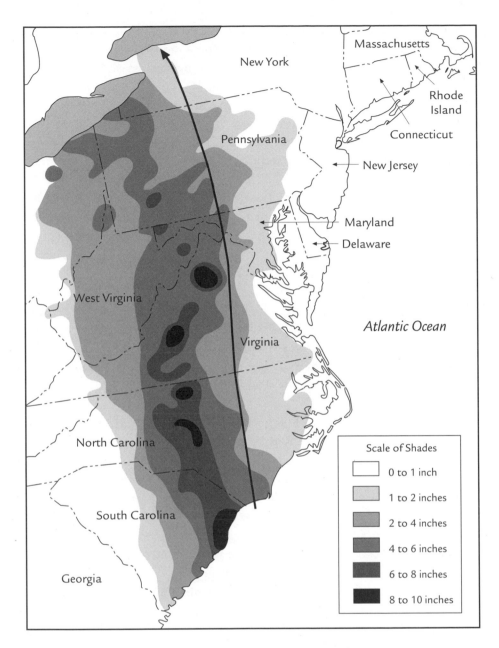

New York

Massachusetts

Pennsylvania

Rhode Island

Connecticut

New Jersey

Maryland

Delaware

West Virginia

Atlantic Ocean

Virginia

North Carolina

Scale of Shades

0 to 1 inch

1 to 2 inches

2 to 4 inches

4 to 6 inches

6 to 8 inches

8 to 10 inches

South Carolina

Georgia

Forests of pine and other trees appeared to be scorched by fire. All trees and plants along the coast appeared to have been burned. Groves of pecan trees were heavily damaged. Building destruction is said to have been greater in the interior sections of North Carolina than at inland points near the coast. From all indications, the hurricane did not decrease in intensity as it moved into the interior.

Inland, out of reach of the rising waters, a tremendous area of North Carolina received heavy damage from high winds. An estimated one-third of all buildings east of the 80th parallel received some damage. Roofs and television aerials were the most widely hit but some radio towers, outdoor theaters, and many signboards were counted among the losses.

Heavy rains fell from northeastern South Carolina to central North Carolina and Virginia. The areas that recorded the greatest rainfall amounts were all on the western side of the storm track, and many eastern locations received as little as one inch. At least ten stations in North Carolina established new twenty-four-hour records for rainfall. These records included about 6.5 inches for Burlington, High Point, and Lexington and 9.72 inches in Carthage, located in the sandhills area of the state. A U.S. Geological Survey rain gauge at Robbins, several miles north of Carthage, measured 11.25 inches. Several locations in northern Virginia recorded more than 10 inches, and new rainfall records were established all along Hazel's northern course.

Although Hazel's most awesome destruction occurred in eastern North Carolina, its inland track through the mid-Atlantic and northern states was unexpectedly intense. After passing Goldsboro and Raleigh at about 1:30 P.M. on October 15, the storm passed through Warren County and into Virginia at about 2:30 that afternoon. At that time, the storm's barometric pressure had only risen to 28.50 inches, up less than one inch from its recorded low near Cape Fear of 27.70.

Hazel accelerated significantly as it passed through Virginia, attaining a forward speed of almost 50 mph. Within four hours, the hurricane had passed over the state, dumping torrential rains and unleashing winds gusting to 100 mph. The storm center passed just west of Richmond, yet winds there were some 30 mph less than in Norfolk, which was on the eastern edge of the storm.

Damages in Virginia were extensive from both winds and tides. Several ships in the James River were either sunk or wrecked, including the battleship *Kentucky*, which broke its moorings and ran aground some one thousand feet from its berth. In all, thirteen Virginians were killed during Hazel, and damages for the state were conservatively estimated at $15 million.

Power failures were commonplace as the storm churned its way through Maryland, the District of Columbia, and Pennsylvania. Wind gusts of 98 mph were reported in Washington, D.C., at 5:07 P.M., and strong winds brought down trees, power lines, and radio towers throughout the region. Winds in Baltimore reached 84 mph, and a six-foot tide flooded streets and basements near Baltimore harbor. Wilmington, Delaware, reported gusts to 98 mph, and Philadelphia recorded gusts to 94 mph. Property damages mounted as Hazel sped relentlessly to the north.

As heavy rains poured out across western Pennsylvania, flash floods swept away cars and homes, and twenty-six lives were lost. The death toll throughout Hazel's northern course was already high, as twenty-two had perished in Virginia, Maryland, Delaware, and the District of Columbia. Then, as the storm raced through western New York state, more tragedies occurred. There, twenty-one deaths were attributed to the hurricane, of which five were caused by falling objects, five by electric shock, four by automobile accidents, three by falls, and four by other causes.

Very few hurricanes in recorded history have maintained their vigor the way Hazel did as it blasted through the interior sections of the northeastern states. As the storm passed through the western counties of New York, winds near 100 mph continued to uproot trees, peel back roofs, and snarl power lines. Then, at approximately 10:00 P.M., just twelve hours after the hurricane had made landfall on the Carolina coast, Hazel joined forces with a weak low-pressure system near Buffalo. Extremely heavy rain fell on numerous communities in western New York, and some city streets were flooded by almost two feet of water. Many highways and bridges were washed out, and the effects of wind and water once again intensified.

As Hazel crossed Lake Ontario, it carried its rampage into Canada. Several locations reported wind gusts of 110 mph. More than seven inches of rain turned the Humber River, which flows along the western edge of Toronto, into a torrent, washing away homes and automobiles. Scores of victims were trapped in the flash flood that swept through the Toronto area. Here, the tragedies continued, as seventy-eight more lives were lost and property damages were estimated at $100 million.

Hazel was last seen crossing the Arctic Circle on its way to Scandinavia, where it eventually fell apart. This incredible hurricane had not lost its momentum and faded away as many storms do when they move over land. Instead, Hazel's destruction was spread over two thousand miles on its northward trek from the Caribbean through North America.

In North Carolina, the destruction left by Hazel was likened to the battlefields of Europe after World War II. Evidence of the storm's violent winds stretched across the state, leaving residents with the task of cleaning up virtually every city street and country road in the eastern half of the state. And the storm tide that swept over the Brunswick and New Hanover beaches brought massive destruction to the coast and was, by all accounts, unparalleled in Tar Heel history.

When Hazel made landfall at Little River, near the North Carolina–South Carolina state line, its deadly storm surge and intense winds reached their peak. At Myrtle Beach, hurricane-force winds began around 6:00 A.M. and continued to intensify until the eye reached the coast at 9:20 A.M. The South Carolina beaches were battered by northeast winds estimated at 130 mph and waves that crested at thirty feet. Hardest hit were locations near the point of landfall, which included Garden City, North Myrtle, Windy Hill, Cherry Grove,

Wreckage filled the waterfront streets of Southport after hurricane Hazel. (Photo by Art Newton; courtesy of the State Port Pilot)

Debris was piled high around the amusement center at Ocean Isle. The Brunswick County beaches suffered the most extensive damages from Hazel. (Photo courtesy of the Cape Fear Museum)

and Ocean Drive. Throughout this stretch of coastline, storm-surge levels ranged between fourteen and seventeen feet above mean low water. Although the destruction in South Carolina was greatest on the northern beaches, homes and piers were damaged as far south as Georgetown. Hazel killed one person in South Carolina and brought $27 million in property damages to the state.

Across the border in North Carolina, the south-facing beaches of Brunswick County caught the brunt of Hazel's fury. Robinson Beach, Ocean Isle, Holden Beach, and Long Beach were hardest hit; virtually every home or structure was washed away or severely damaged. In most cases, oceanfront

The National Guard was posted along the beachfront following hurricane Hazel to guard against looting and to restrict sightseers. (Photo courtesy of the Cape Fear Museum)

At Long Beach, most of the cottages that once lined the strand were destroyed by Hazel's seventeen-foot storm surge. Many were washed over into marsh thickets on the leeward side of the island. In the weeks following the storm, some owners hired crews to salvage their cottages by pulling them out of the marsh and repositioning them on the beach. (Photo courtesy of Hugh Morton)

cottages were first battered by waves and 150-mph winds, then swept off their foundations and tossed several hundred yards into marsh thickets. Those unlucky souls who were caught in Hazel's rapidly rising storm tide either rode out the upheaval of their homes or perished in the ordeal.

After surveying the Brunswick beaches in the aftermath of Hazel, the Weather Bureau office in Raleigh issued the following report:

All traces of civilization on that portion of the immediate waterfront between the state line and Cape Fear were practically annihilated. Grass-covered dunes some 10 to 20 feet high along and behind which beach homes had been built in a continuous line 5 miles long simply disappeared, dunes, houses, and all. The paved roadway along which the houses were built was partially washed away, partially buried beneath several feet of sand. The greater part of the material from which the houses had been built was washed from one to two hundred yards back into the edge of low-lying woods which cover the leeward side of the islands. Some of this material is identifiable as having been parts of houses, but the greater portion of it is ground to unrecognizable splinters and bits of masonry. Of the 357 buildings which existed on Long Beach, 352 were totally destroyed and the other five damaged. Similar conditions prevail on Holden Beach, Ocean Isle, Robinson Beach, and Colonial Beach. In most cases it is impossible to tell where the buildings stood. Where grassy dunes once stood there is now only flat, white, sandy beach.

The *State Port Pilot* reported the story of two survivors who endured the worst of the storm: Connie and Jerry Helms. The couple had come to Long Beach for their honeymoon and were out roller-skating the night before Hazel struck. The Coast Guard warned other residents of the hurricane's approach, but the newlyweds arrived back at the beach too late to hear the warnings. They went to bed with no knowledge of the ominous storm poised off the coast.

At dawn the next morning, the Helmses awoke to the sound of pounding surf and chairs blowing about their porch. Huge waves began crashing over the dunes as the couple scrambled to leave their cottage. With no time to even pack their clothes, they jumped into their car, but it wouldn't start. They ran to their jeep, but it wouldn't start either. By this time, the surging ocean water was waist deep in the street, and they knew it was becoming deeper and rougher by the minute. They made their way to a nearby two-story house and broke through a door to get inside. As the tides continued to rise, the second floor buckled and the couple feared they might perish if the house collapsed.

"We started seeing houses exploding then floating away," Jerry Helms recalled. "Sometimes you could see the whole house flying through the air. There was a little cinderblock house next door, and a breaker went over top of the house, and after it went over it was gone. Stoves, refrigerators, and houses were flying through the air. Before we got out of the house we were in, we saw a guy standing in the doorway of another house floating by. He was found the next day buried in the sand."

The couple watched as the entire island went underwater and every house within sight was swept away or pounded to splinters by massive waves. Large timbers, roof sections, and appliances of every kind whisked past the cottage where they had taken refuge. Then, as the water swirled just inches below the second-story windows, the Helmses knew that their shelter was about to give in to the storm's waves and winds. Tying themselves together with a flannel blanket, the newlyweds escaped the collapsing house through a high window. Connie Helms, who could not swim, climbed onto a cotton mattress while her husband jumped into the raging waters.

"We went right out to sea on that mattress," Jerry recalled. "By that time, all the two-story houses were covered by water, and most of the other stuff was already gone. But we never had a chance to think about what was happening or be scared. We were just trying to hold on to that blanket and stay alive."

The currents did not take the couple out to sea but instead carried them toward the leeward side of the island. To stabilize their makeshift life raft, Jerry grabbed a section of a house that drifted by and positioned their mattress on top of it. A short time later, their raft became lodged in some treetops near Davis Creek. There the couple endured the storm for several more hours.

"It got real calm when the eye passed over, and then it started raining real hard. It felt like little bullets," Jerry recalled. "There was a lot of times we thought we were gone. The only thing that saved us was the good Lord and that blanket."

When the hurricane winds subsided and the floodwaters receded, the Helmses climbed down from the treetops. They then made their way across Davis Canal — still clutching the blanket that had helped save their lives. The couple later found their way to Southport, where they visited Dosher Memorial Hospital for treatment. Later that night they went back to their home in Whiteville.

The following day, Jerry Helms returned to Long Beach to search for the remains of the car, the jeep, and the cottage. Both vehicles were buried deep in the sand, and the only remaining fragment of the house was one corner, held in place by a single post. About a mile up the beach, he found his refrigerator, still intact, with its door sealed. "The drinks in the refrigerator were still cold and the only thing broken was a bowl of peas, so we sat down and ate lunch out of it," he said. The honeymoon was over for the Helmses, but their incredible struggle for survival was not soon forgotten.

Several miles west of Long Beach, the resort of Ocean Isle was the scene of the single greatest tragedy Hazel brought to the Carolinas. The beach was connected to the mainland by a small ferry that was destroyed in the early stages of the storm. Given no means of escape, a group of eleven people gathered in one of the island's larger buildings to rise above the oncoming storm surge. As the winds and waves increased, the building began to crumble, forcing the group to swim through the breakers to a nearby truck. With the women and children inside, the men tried in vain to keep the vehicle upright as the tides continued to rise. Tragically, their efforts were no match for Hazel's fury and the truck was soon washed away. Of the eleven, only two survived. Thirty-three cottages that had stood on Ocean Isle were completely de-

At Southport, Hazel's storm tide lifted thirty-five-ton shrimp trawlers over their docks and into the streets. (Photo by Art Newton; courtesy of the State Port Pilot*)*

At Southport, Hazel's storm tide lifted thirty-five-ton shrimp trawlers over their docks and into the streets. (Photo by Art Newton; courtesy of the State Port Pilot*)*

Debris covered streets and yards throughout downtown Southport. (Photo by Art Newton; courtesy of the State Port Pilot*)*

stroyed. Only two houses remained, and they were washed almost one mile across the island into the marsh.

In Southport, Hazel's storm tide forced ocean water through the mouth of the Cape Fear River, flooding the waterfront streets with a surge of eight feet. All twenty of the town's shrimp houses and fuel docks were destroyed. Restaurants along the river and several warehouses were demolished. The storm tide lifted thirty-five-ton shrimp trawlers over the seawall and swept them into town, crushing cars and homes along the way. Along Bay Street, majestic old oak trees were toppled by winds near 140 mph. For several blocks near the waterfront, large piles of debris choked the town's streets.

As hurricane Hazel came ashore on the morning of October 15, 1954, the people of Southport scrambled to avoid the storm's rapidly rising tide. The surge wrecked the many shrimp houses that lined the riverfront and carried the remaining debris far into the town's streets. (Photo by Art Newton; courtesy of the State Port Pilot)

The Southport waterfront after Hazel. (Photo by Art Newton; courtesy of the State Port Pilot)

Lewis Hardee owned two trawlers and a shrimp-packing house on the Southport waterfront. As the storm reached its peak, his boats floated over the docks and his shrimp house was destroyed. Like many others in the small town, Hardee made his living on the water. "Back then, nobody had insurance on boats in Southport," Hardee recalled. "Things were a mess. Boats were scattered all in the marsh from the Intercoastal [waterway] to the Coast Guard Station. We all had a warning, but nobody knew it would be like it was."

Hardee had just rebuilt his shrimp house and dock the summer before

Portions of Caswell Beach were swept clean by Hazel's storm tide, which broke apart the only road down the beach. (Photo courtesy of Mrs. J. T. Barnes)

Southport after hurricane Hazel. (Photo by Art Newton; courtesy of the State Port Pilot)

Hazel arrived and tore them apart. A fire the previous year had damaged the old structures, and his new pier and house had been built with costly cypress. Hazel's powerful winds and waves broke them to pieces and scattered their timbers along the shore. "After the storm, I took a carpenter's crayon and went along the waterfront and put my name on every board," Hardee remembers. "You know they weren't hardly seasoned, they were new lumber, just a year old. All two-inch cypress decking. With that lumber I built another new fish house down at the yacht basin."

After Hazel passed, the fishermen of Southport faced the challenge of removing their large trawlers from high ground, repairing their damaged hulls, and returning them to the water. Fortunately, a contractor at the nearby ammunition facility at Sunny Point offered assistance in this effort. A crew from Diamond Construction Company brought in an eighty-five-foot crane that lifted the large boats from the streets and yards of Southport and placed them back in the water. This task took several days, but the residents and fishermen of this storm-ravaged town were grateful for the help.

Farther up the river in Wilmington, flooding in the Cape Fear reached its highest level in recorded history. Floodwaters damaged numerous warehouses along the waterfront, and dozens of cars were submerged by the rising water. Overall damage to the city was not extensive and was limited mostly to broken storefront windows and uprooted trees. High winds snapped telephone poles in several locations, leaving more than half the area's residents without any means of communication. Wilmington was without electricity for three days. Two thousand residents were sheltered in twenty Red Cross evacuation centers around the city.

Southport's yacht basin before Hazel's peak. The trawlers pictured here eventually floated over their docks and drifted down Moore Street, where they came to rest in backyards and side streets. (Photo by Art Newton; courtesy of the State Port Pilot)

Diamond Construction Company came to the aid of Southport residents by delivering a large crane to remove wayward boats from the town's streets. (Photo courtesy of Leila H. Pigott)

The Wilmington Reserve fleet in the Brunswick River basin suffered about $1 million in damage from the storm. Several of the large ships broke their mooring lines when high winds "snapped cable an inch and five-eights in diameter like it was ribbon." Three liberty ships drifted upriver with the winds and threatened to wreck the Brunswick River bridge. At the peak of the storm, a tugboat captain made a heroic effort to stop the ships just one hundred yards from the bridge. His courageous efforts saved the concrete span that provided the only link between Brunswick and New Hanover Counties.

Carolina Beach was hit hard by Hazel, largely because the popular resort found itself in the hurricane's powerful northeast section. Property damage

there totaled $17 million, much more than in any other coastal community. The heaviest damages occurred along the oceanfront, where the storm's tidal surge bashed homes and the downtown amusement area. In all, 362 buildings were completely destroyed and another 288 suffered major damage.

A number of residents stayed on Carolina Beach through the hurricane and witnessed Hazel's attack on their town. They saw large waves roll through Mack's 5 & 10 store and watched as the steel pier collapsed and sank. Dozens of houses floated off their foundations and "crashed together like bumper cars." One man was crouched in his living room when a surging wave heaved an eight-by-eight-inch timber through four walls of his home. After the hurricane passed, the giant piece of lumber had to be cut into four pieces to be removed.

According to a story by Susan Gerdes that appeared in the *Tidewater*, eighty-year-old Alex McEachern refused to leave his Carolina Beach home upon Hazel's approach. To escape the rising waters, he and his dog climbed atop a freezer in his pantry to ride out the storm. Miraculously, even though Mc-

Deep sand filled the streets of Carolina Beach after hurricane Hazel. National Guard troops patrolled the area to search for storm victims and to protect homes and businesses from looting. (Photo courtesy of the Cape Fear Museum)

Alice Strickland, town clerk of Carolina Beach, remained in the small town hall during hurricane Hazel. Undaunted, she cheerfully directs workers while the receding storm tide swirls around her knees. (From Making a Difference in North Carolina, by Ed Rankin and Hugh Morton; photo courtesy of Hugh Morton)

Eachern's house was torn apart by the wind and waves, the pantry was unharmed; in fact it was the only room left standing. After the storm subsided, the lucky twosome climbed down from their refuge and found their way into town.

During the early part of the storm, emergency workers used the Carolina Beach Town Hall auditorium as disaster headquarters. But as the floodwaters rose, town officials packed up their equipment and important documents and retreated to the China Cafe, which stood on slightly higher ground. Ultimately, fourteen blocks of the town were underwater. One house caught fire

The Breakers Hotel, Wilmington Beach. (Photo courtesy of the Cape Fear Museum)

A sentry stands guard in downtown Carolina Beach. There were reports in the days following Hazel that looters were launching boats and even swimming over to the storm-ravaged resort communities of Carolina and Wrightsville Beaches. (Photo courtesy of the Cape Fear Museum)

Carolina Beach during hurricane Hazel. (Photo courtesy of Hugh Morton)

during the storm, but firefighters were unable to reach it because of the flood. As officials looked on, the residence burned down to the waterline, which was about four feet above the ground.

Up the coast at Wrightsville Beach, Hazel's storm surge ruined dozens of large oceanfront cottages. Estimated at twelve feet above mean low water, the

The first row of cottages facing the ocean at Wrightsville Beach was eliminated by Hazel's storm surge. (Photo courtesy of Hugh Morton)

tide swept over the island, destroying 89 buildings and severely damaging another 155. Property damages were estimated at $7 million, including severe destruction to the town's sewage plant.

When Hazel struck the Wilmington area, only twelve state troopers were on duty, but additional support was brought in quickly. By the following day, fifty troopers from as far away as Asheville had come to assist local police with the around-the-clock task of maintaining checkpoints, directing traffic, and assisting those in need. Numerous accounts of heroic rescues were compiled after the storm, including several rescues made on foot in chest-deep water. Many local officers were said to have faithfully maintained their duties while their own homes were demolished in the storm.

Checkpoints were set up in key locations to prevent immediate access to the hard-hit barrier islands. Thousands of eager sightseers traveled to the coast to witness the destruction, but most were turned away. Looting was a severe

Wrightsville Beach after hurricane Hazel. (Photo courtesy of Hugh Morton)

Wrightsville Beach. (Photo courtesy of the Lower Cape Fear Historical Society)

problem, and on some beaches the National Guard was brought in to assist the police with protecting property. According to local news reports, some looters launched small boats and even swam across the waterways to evade police and made off with radios, stoves, refrigerators, and cash.

Even along the coastline north of Wrightsville there was considerable destruction. At New Topsail Beach, 210 of the 230 houses were destroyed and property damages were estimated at $2.5 million. The Topsail Island drawbridge was "carried away" by Hazel's furious winds and tides, and a Marine Corps amphibious vehicle was the only means of transport to the island for days. Snead's Ferry and Swansboro suffered from extreme high tides, which

Up the coast at Topsail Island, Hazel's storm surge was deadly to the cottages that lined the beach. (Photo courtesy of the Cape Fear Museum)

Hurricane Hazel washed out numerous bridges throughout eastern North Carolina, including this one at Snead's Ferry. (Photo courtesy of the Cape Fear Museum)

deposited fishing boats and pleasure craft high and dry in streets and back-yards.

In Carteret County, 120 miles north of where Hazel made landfall, residents witnessed the worst hurricane in many years. Hundreds of citizens took refuge in Morehead City Town Hall and the county courthouse in Beaufort. Damage was extensive in several locations near the water, including Atlantic Beach and the causeway that joins Morehead City and Beaufort. Property damages were reported at $2 million, but fortunately there were no reports of deaths or serious injuries. Thirty-five homes were destroyed, and scores more suffered minor damage.

Even though Hazel's eye passed nearly a hundred miles to the west, Carteret County was within reach of the storm's violent right-front quadrant. High tides flooded parts of Morehead City, as seen in this photo, which was taken looking westward from the old Atlantic Beach bridge. (Photo courtesy Clifton Guthrie)

In Atlantic Beach, Hazel's storm surge pounded the boardwalk area to rubble. Twenty-foot waves washed away a section of the Atlantic Beach Hotel. On the other end of the boardwalk, waves washed through the lobby of the Ocean King Hotel, undermining the structure. The Triple S fishing pier was battered by the surf and its tackle shop was badly damaged. After the storm, only two hundred of the pier's one thousand feet remained. The old beach highway that connected Salter Path and Emerald Isle was swept away in two places. In these areas, Bogue Sound connected with the ocean, but only briefly, until heavy equipment was brought in to fill in the overwash.

In Morehead City, a fish house, dwelling, and skating rink were washed from their foundations. Large glass windows were smashed and trees were

An eight-foot storm surge battered the Pavilion at Atlantic Beach. (Photo courtesy of the Carteret County News-Times)

Front Street in Beaufort was under several feet of water during hurricane Hazel. At the peak of the storm, water nearly covered the parking meters shown here. (Photo courtesy of the Carteret County News-Times)

toppled by winds that gusted to near 100 mph. At the peak of the storm, surging waters flooded the basement of the Morehead City hospital. For hours, fire trucks pumped the water out so that basement facilities could continue to be used. Nurses in the basement worked with water up to their knees, scrambling to remove patients and save equipment.

Tony Seamon of Morehead City witnessed the effect Hazel had on the county. He and his father ventured downtown during the storm. "Daddy wanted to check on the restaurant," Seamon recalled. "We drove down Twentieth Street and the wind whipped the water across the road. We came to the

Oceanfront destruction in Atlantic Beach after Hazel.

old wooden picket bridge and it was submerged. The only way we could drive through that water was to open both car doors and let the water come on through."

As they reached downtown Morehead, Seamon's dad asked him to investigate the damage to the restaurant. "He told me, 'You got to get out and look inside,'" Seamon remembered. After one failed attempt to wade through the waist-deep water, he tried again to make it to the Sanitary Restaurant. "The current through the street was like a river. I eventually went under swimming to avoid the wind. When I got to the restaurant, I could look through the door and see all the tables and chairs floating around. Daddy had made the decision to cut holes in the floor to equalize the pressure. At least the whole place didn't float away."

After the hurricane passed, the Sanitary Restaurant was cleaned up and put into service as a feeding center. The Red Cross set up generators to provide power, food was brought in, and coupons were issued for meals. Work crews involved in cleanup efforts were fed around the clock until power was restored to the area and things returned to "normal."

Other portions of Carteret County were hit hard by the storm. Several homes along the Morehead-Beaufort Causeway were totally wrecked, and their debris was piled high in the middle of Highway 70. Huge waves rolled across the causeway for hours, washing away homes and depositing small skiffs and one large cruiser in the roadway. Several cars were abandoned by their owners on the east side of the Beaufort bridge when the unwary motorists became trapped by the hurricane's surging tide. The cars choked when

In Raleigh, as in many other cities and towns across eastern North Carolina, residents survived without electricity in the days following hurricane Hazel by cooking meals over campfires. (Photo courtesy of the News and Observer Publishing Co./N.C. Division of Archives and History)

Hazel's violent winds lashed the state, affecting inland areas as well as the coast. Warehouses and other buildings in Wilson were toppled by gusts near 100 mph. (Photo courtesy of Raines and Cox Photographers)

water completely covered their engines, and the drivers were forced to wade through the tide to safety.

In Beaufort, the downtown businesses along Front Street suffered heavy losses. Every store was flooded with seawater, which covered the entire street to a depth of three feet. Gusting winds caused plate-glass windows in numerous stores to shatter, resulting in even greater damage from wind-driven rain. City Appliance Company, Bell's Drug Store, and Merrill's Men's Store were among the hardest hit. Numerous small boats were washed into the streets, and at least four cabin cruisers sank in Taylor's Creek. The highest water in Beaufort occurred at 11:15 A.M., when the barometer fell to its low point (for

Hurricane Connie damaged numerous cottages in Atlantic Beach, many of which had only recently been repaired following hurricane Hazel the previous year. (Photo courtesy of Robert Lewis)

Carteret County) of 29.06 inches. The tides in Morehead reached their peak an hour later, around noon.

Across the county, Hazel battered homes, boats, and utilities. Many houses that had just been reshingled after hurricane Edna were stripped again by gusts of 90 to 100 mph. Power lines and poles were tangled and communications were cut. A section of the North River Bridge was washed away, but rapid work by state highway crews had the bridge repaired just five days after the storm.

Although the Pamlico and Albemarle Sounds region was on the eastern edge of Hazel's course through the state, numerous communities throughout the area suffered serious flooding. The Outer Banks north of Ocracoke were not severely affected, but cities such as Washington and Belhaven saw extensive damage from the rising tides. New Bern, Edenton, and Elizabeth City also reported flooding. But across a wide portion of inland North Carolina, far from the effects of the tides, Hazel's powerful winds brought significant destruction to more than two dozen counties.

Although most of Hazel's fatalities in North Carolina occurred along the beaches of Brunswick County, several deaths resulted from the harrowing winds that raced across the state. In Wallace, a tobacco warehouse collapsed, crushing a Warsaw man underneath the rubble. In Parkton, a one-month-old infant was killed when a large tree fell through her home. The child's mother

Hurricane Connie was followed so closely by hurricane Diane that the floods affecting much of eastern North Carolina had little time to subside. Sounds and rivers like the Pamlico spread far beyond their banks. (Photo courtesy of Roy Hardee)

was seriously injured, as she had been lying next to her daughter in bed when the massive tree came crashing down. Other deaths occurred as a result of electrocution, falls, and automobile accidents. Across North Carolina, hospital emergency rooms were filled with victims of the storm. At least two hundred injuries were reported, some serious, in the six hours Hazel visited the state.

By most accounts, it was the most destructive hurricane in Tar Heel history. In North Carolina the toll was heavy: nineteen people killed and over two hundred injured; fifteen thousand homes and structures destroyed; thirty-nine thousand structures damaged; thirty counties with major damage; and an estimated $136 million in property losses. But when the hurricane's effects in North Carolina are combined with those of the other states, as well as with those of Canada and Haiti, the numbers climb: over six hundred dead and an estimated $350 million in property damage. And of course, the damages are in 1954 dollars.

The great hurricane of October 1954 became a benchmark in the lives of many North Carolinians who endured the storm. From Holden Beach to Henderson and everywhere in between, anytime the topic of hurricanes is raised, stories about Hazel are sure to follow. Stories of heroic rescues and tragic losses are well remembered, as are testimonials to the awesome destructive forces

the storm displayed. Hazel ranks as one of the most catastrophic hurricanes to strike the United States in the twentieth century. Fortunately, storms of its magnitude are relatively rare events, and few other hurricanes deserve comparison with it. But much to the dismay of the people of eastern North Carolina, the active 1954 hurricane season that had spawned Carol, Edna, and Hazel was merely a prelude to the 1955 season, when three more storms would strike the state.

CONNIE (AUGUST 12, 1955)

Less than one year after Hazel carved its way through the Carolinas, residents along the Tar Heel coast braced themselves for the approach of another serious storm—hurricane Connie. As it turned out, however, Connie was only the first of three hurricanes to make landfall in North Carolina during the 1955 season. Hurricane Diane followed closely on Connie's heels, and hurricane Ione arrived a month later. All three storms spun across the state in one frightening six-week period, between August 12 and September 20. These three events brought unprecedented destruction to the eastern counties and compounded the earlier damages caused by Hazel.

As hurricane Connie passed north of Puerto Rico on August 6, reconnaissance crews measured its barometric pressure at 27.88 inches and estimated its maximum winds to be near 145 mph. By August 8, Connie's forward movement had slowed considerably, and the storm wobbled erratically as it approached the Carolinas. By the time it crossed over Cape Lookout on the morning of August 12, Connie's intensity had decreased considerably. After

passing through the Pamlico and Albemarle Sounds region, it continued
northward up Chesapeake Bay before turning to the northwest. On August
14, Connie dissipated somewhere over Lake Huron.

As Connie moved across the North Carolina coastline its winds were barely
of hurricane force. Wilmington reported sustained winds of 72 mph with
gusts of 83 mph. Winds of 75 mph were reported at Fort Macon, with gusts
that reached 100 mph. Wind damages were relatively light along the coast, although some television aerials and roof shingles that had been repaired after
Hazel were lost.

Before Connie moved ashore in Carteret County, and while it was still staggering about in the Atlantic, several tornadoes were reported in the Carolinas.
Early in the evening on August 10, five twisters touched down in South Carolina, from Georgetown northward. One was reported near Penderlea, North
Carolina, where five buildings were demolished and one person was injured.
These tornadoes were most likely spawned in the broad spiral rainbands that
extended from the center of hurricane Connie.

Although overall wind destruction from Connie was light, torrential rains
and prolonged high tides brought extensive flooding to the North Carolina
coast. For several days the storm moved sluggishly, and broad-scale winds
over the Atlantic pushed a high mound of water toward the shore. Connie's
slow movement across the state aggravated the situation as tremendous rains
were dumped on the eastern counties. The result was that thousands of acres
of farmland were flooded by a combination of extreme high tides and heavy
runoff.

Huge waves pounded the coastline, and beach erosion was said to have been

Hurricane Diane made landfall just five days after hurricane Connie in August 1955. Although Diane was a category-one storm when it hit, it went on to become one of our nation's most tragic hurricanes. (Photo courtesy of the News and Observer Publishing Co./N.C. Division of Archives and History)

The National Guard was available in Belhaven to evacuate residents in need during hurricane Diane. (Photo courtesy of Roy Hardee)

worse than that caused by Hazel. Tides were about seven feet above normal on the beaches from Southport to Nags Head, and flooding in the sounds and near the mouths of rivers was estimated to range from five to eight feet.

At Morehead City, near where Connie's irregular eye crossed the coastline, almost 12 inches of rain were recorded. Heavy rains fell throughout the mid-Atlantic states as the storm dragged slowly northward. Amounts in excess of 9 inches were measured in eastern Pennsylvania, New Jersey, and Delaware. At La Guardia Field in New York, 12.20 inches of rain were recorded in a thirty-eight-hour period on August 12–13.

DIANE (AUGUST 17, 1955)

Although the heavy rains from Connie did not bring extensive damages to North Carolina or the Northeast, they did saturate the ground, overfill rivers and streams, and set the stage for the record-breaking floods of hurricane Diane. Diane followed so closely behind Connie that coastal residents had little time to prepare for its approach. And no one was prepared for the deadly flash floods that swept through the Northeast.

Because the two hurricanes hit just days apart, officials were unable to as-

Diane forced some residents to move to higher ground. (Photo courtesy of Roy Hardee)

sess the damages from each separately, and many records were combined. Connie and Diane delivered a one-two punch to a broad region, as floods from Diane were amplified by the soaked soils from the previous storm. Diane's heavy rains brought flash floods to several states. These floods killed two hundred and established a new record in destruction—Diane was declared by some to be the first "billion-dollar hurricane."

The very day Connie raked over the North Carolina coastline, Diane reached hurricane intensity some one thousand miles at sea. For a while, Diane turned toward the open ocean, possibly influenced by the movements of Connie. But then the storm veered to the northwest and curved toward the North Carolina coast. In the early morning hours of August 17, just five days after Connie's arrival, hurricane Diane moved inland over Carolina Beach.

Ship reports had indicated that Diane's winds were near 125 mph while it was at sea. But fortunately, and as with Connie, its intensity dropped as it approached land. Winds of 50 mph were reported at Cape Hatteras, and a gust of 74 mph was recorded in Wilmington. These gusts had a minimal effect on the coastal region, and wind-damage reports were primarily from inland counties that suffered crop damage.

But as with Connie, Diane's wind-driven tides and torrential rains brought extensive flooding to the Tar Heel coast. Tides associated with Diane were generally higher than those of the previous hurricane and ranged from five to nine feet above mean low water along the beaches. Some coastal rivers and sounds also crested at nine feet above normal. In Belhaven, the water rose three feet above floor level in the business district. In some portions of downtown New Bern and Washington water was "waist deep." Flooding in the Cape Fear River was extreme, and damages were reported along the Wilmington city docks. In all, one thousand people were evacuated from the low-lying areas and towns near the water's edge.

In Carteret County, Diane had similar effects. Residents were still reeling from Connie when the warnings were issued that a second storm was approaching. Some shop owners in Beaufort lined their doors with sandbags in hopes that rising waters might be kept at bay. High tides brought flooding to most causeways and bridges, and the North River Bridge was once again put "out" by the storm. The motor on the Atlantic Beach drawbridge burned out just prior to Diane's arrival, and prisoners from nearby Newport Prison were brought in to turn the steel span by hand. Twenty-three homes at Salter Path were damaged by high water, but artificial sand dunes hastily built after Connie helped limit the destruction to many beachfront homes.

But along many southern beaches and as far north as Cape Hatteras, beach erosion was severe. Diane's slow movement to the northwest caused prolonged winds to push salt water out of Pamlico Sound and into the farms and fields of the east. Thousands of acres were flooded again, just as they had been during Connie. Supplementing this tidal flood, Diane dumped heavy rains across the state. But although North Carolina was inundated by the combined effects of the two August hurricanes, the flooding and rainfall that followed in the Northeast delivered the catastrophic destruction for which Diane is remembered.

After it came ashore on the beaches of New Hanover County, Diane continued on a steady inland track. By midafternoon on August 17, the storm center had passed over the Raleigh-Durham Airport and was beginning to curve gently into Virginia. By the early morning hours of the eighteenth, it had begun to turn toward the northeast, and it passed near Lynchburg, Virginia, on its way through Philadelphia, Atlantic City, and Nantucket, Massachusetts. While the storm tracked northward through Virginia, winds diminished to only 35 mph. Diane's winds were of little consequence when compared with the record-breaking rains that were unleashed as the hurricane drifted to the northeast. Several locations along the eastern slopes of the Blue Ridge Mountains received in excess of ten inches, and portions of eastern New York and southern New England recorded over twelve inches within a twenty-four-hour period. A report from the National Weather Service detailed some of the destruction: "The excessive rains of hurricane Diane fell on many of the same localities previously drenched by rains associated with hurricane Connie, and

Governor Luther H. Hodges examines a tracking chart to study the path of the next approaching hurricane, Ione, in September 1955. (Photo courtesy of the News and Observer Publishing Co./ N.C. Division of Archives and History)

combined they produced the most devastating floods the valleys of eastern Pennsylvania, northwestern New Jersey, southeastern New York and southern New England have ever experienced. In the vicinity of Stroudsburg, Pa., 75 persons lost their lives when rapidly rising water from Broadhead Creek swept away a summer camp. The floods in Virginia and portions of North Carolina and West Virginia were damaging in some localities, but less severe than those in the northeastern states."

By late in the afternoon on August 19, Diane was on its way back to sea, where it died quietly over the North Atlantic. It is estimated that the storm claimed close to two hundred lives during its two-day visit, and almost all were lost in the deadly flash floods that washed through the Northeast. Initial damage estimates exceeded $800 million, surpassing all other hurricanes in U.S. history. Ironically, the vast majority of the destruction occurred not along the coastline but in the inland sections of several states.

In North Carolina, no lives were lost during either Connie or Diane, but damages from the two storms exceeded $80 million. Twenty-two counties were affected by the hurricanes' winds and water, and numerous construction projects still under repair from previous storms were undermined. Connie had registered as a category-three hurricane when it hit, Diane as a category two. Neither storm matched Hazel in intensity or local destruction, but many

North Carolinians were beginning to wonder what wrath had fallen upon them. In churches across the state, special "hurricane prayer" services were held to comfort victims and build hope for more peaceful times.

IONE (SEPTEMBER 19, 1955)

The 1955 hurricane season in North Carolina did not end with Connie and Diane. One month after Diane's departure, hurricane Ione struck the coast near Salter Path, a few miles west of Atlantic Beach. Ione became the third significant hurricane to make landfall that summer, something of a rare occurrence in North Carolina. Ione eventually inundated record portions of the coastal plain and established new high-water marks in numerous locations. Several new records for rainfall were also recorded across the state.

Like many September hurricanes, Ione was born out of an easterly wave that passed through the Cape Verde Islands on September 6. By the time it reached the western Atlantic and was in position to threaten the Carolina coast, Ione was a major storm. On September 17, it reached peak intensity with sustained winds of 125 mph and a central pressure of 27.70 inches. When it crossed the coastline at 5:00 A.M. on the nineteenth, the pressure had risen to about 28.00 inches and winds had diminished considerably. Ione's winds gusted at around 100 mph in several locations. The highest winds were measured at Cherry Point, where sustained winds of 75 mph and a gust of 107 mph were recorded. The hurricane continued to weaken as it returned to sea off the Virginia coast.

Incredible amounts of rain were dumped on the eastern counties as Ione dragged its heavy clouds across the region. A report from the National Weather Service described the record flooding:

> As Ione crossed the coastline in the vicinity of Morehead City, hurricane associated rains continued for a long period. Most of the damage caused by hurricane Ione was by flooding to crops. According to the State Climatologist at Raleigh, between August 10 and the approach of Ione, eastern North Carolina had been repeatedly drenched with heavy rains with more than 30 inches falling in the wettest sections. The some 16 inches which fell in the same areas with the passage of Ione brought 45-day precipitation totals to values without precedent in the weather history of North Carolina. In the 42-day period August 11 to September 20 the cooperative field station at Maysville reported a rainfall of 48.90 inches. This unprecedented rainfall, approximately one-third of it falling in about 30 hours with hurricane Ione, produced the heaviest runoff of record on downstream tributaries and coastal creeks in North Carolina. The combination of tide waters from the east and flood water from the west inundated the greatest area of eastern

Flooding in New Bern during hurricane Ione. (Photo courtesy of the New Bern–Craven County Public Library)

North Carolina ever known to have flooded. In New Bern, the depth of water which, according to press reports, was 16-feet above normal, was the greatest of record.

In New Bern, the actual flooding was measured at ten and a half feet above mean low water, still one and a half feet greater than the hurricanes of 1913 or 1933. Forty city blocks were flooded; some streets were covered with so much water that parking meters were completely submerged. Many residents were evacuated from their homes by boat. Water measured six feet deep on East Front Street north of Broad Street. By 9:00 A.M. on September 19, the eye of the storm passed just east of New Bern, and the flooding reached its peak. As many as two thousand New Bern residents were sheltered during the storm.

The flooding was so severe across a broad area that rural roads, city streets, and state highways were impassable. Among the highways closed to traffic were N.C. 24, cut by a washout between Swansboro and Morehead City; N.C. 53 near Burgaw; N.C. 172 near Snead's Ferry; U.S. 264 near Washington in Hyde County; N.C. 33, where a bridge was washed out between Aurora and Chocowinity; N.C. 94 across Lake Mattamuskeet; N.C. 11 near Kinston; U.S. 70 near Newport; and U.S. 158, where damages occurred to Little Creek Bridge.

State highway officials reported "anywhere you went, you went in water." Severe damage occurred to the North River Bridge, east of Beaufort. Just as had happened during hurricane Hazel, Ione's tides swept away several wooden spans of the 975-foot drawbridge, immediately isolating three thousand

Carteret County residents. While repairs were being made, amphibious vehicles were brought in to transport supplies and rescue workers to and from the remote down-east areas. In addition, the state highway commission hired a local boat captain to use his trawler for transport.

Man-made sand dunes that had been erected near Atlantic Beach after Diane helped lessen the impact of Ione's storm surge. Mayor Alfred B. Cooper reported that beachfront destruction was not as severe as it had been in the earlier hurricanes. A few houses were undermined and several lost their porches, but overall oceanfront damage was not severe. All of the fishing piers in the Carteret region remained intact after Ione. Near the Morehead–Atlantic Beach Causeway, however, "Mom and Pop's" fishing pier was destroyed by the tide.

The combined effects of Connie, Diane, and Ione were said to have swept away all the dunes along the twenty-five-mile stretch of beach from Cape Lookout to Drum Inlet. This narrow barrier island was "smooth as an airfield" after the storms had passed. Drum Inlet itself was affected by the massive tides and became unnavigable after Ione, when new sand shoals formed and choked the channel. At Sea Level, Otway, Harker's Island, and Beaufort, homes and streets were flooded. Fishing boats and small craft were tossed ashore and wrecked in typical fashion. Along the Beaufort waterfront, numerous businesses were again flooded and store windows were smashed.

The *News and Observer* reported that in Morehead City, the hurricane "raged all day" and was described as the "worst storm in 35 years." More than three feet of water filled many streets. Fish houses, docks, and other structures were washed about, leaving large piles of wrecked lumber along the shore. Winds of 110 mph toppled trees and ripped away shingles and, in some cases, entire roofs. The Morehead City hospital was flooded again during Ione. The basement filled with water just as it had during Hazel, and this time electrical equipment was damaged. Three women were reported to have been in labor during the peak of the storm, but all fared well.

As Ione moved northward through the Pamlico-Albemarle region it slowed and curved to the northeast, passing near Elizabeth City on its way back to the Atlantic. This track carried the hurricane far enough inland to extend damages westward. In Belhaven, most of the town's residents had to be evacuated in amphibious military "ducks." In Aurora, much of the community was inundated and most residents were forced to leave. In Washington, North Carolina, "two-thirds" of the city was flooded as the Pamlico River and numerous creeks overflowed. In Greenville, Ione was described as "more severe and a longer blow than Hazel." In Bethel, Farmville, Plymouth, Elizabeth City, and numerous other down-east communities, damages were severe, especially to crops. Peanuts, tobacco, cotton, and corn were among the hardest hit.

Along the Outer Banks, beach erosion carved away more protective dunes at

Ocracoke and Hatteras Island. Ione had originally been forecast for landfall between Cape Lookout and Cape Hatteras, and residents on Ocracoke were prepared for high waters that never came. Wooden Coca-Cola crates and cinder blocks were used to elevate refrigerators, stoves, freezers, and furniture. One island resident described the preparation in the *News and Observer*:

> People with second storeys are fortunate. Lots of things are laboriously carried upstairs. The loud cackling from a neighbor's yard tells us that the chickens are being caught and being put up on the top roost, with wire stretched beneath it to keep them there. The ducks resent this treatment; they prefer to be penned outside where the swimming is good. The pet dogs and cats are called to safety.
>
> Flashlight batteries, candles, oil, food are stocked for emergency use; also a good supply of drinking water. Two precautions must be observed: 1) before the storm gets too bad and the salt spray begins to blow through the air, disconnect the cistern or water barrel, and 2) before the tide gets too high, let the water come into the house. Don't try to keep it out or you'll float away.

After Ione paid its visit to eastern North Carolina, it turned slowly back to sea, sparing the northeastern states a repeat of the destruction brought on by hurricane Diane. Ione's wobbling, erratic movements were described by officials from the Weather Bureau as "a forecaster's nightmare." Although they had done their best to outguess the storm's next moves, its return to sea was not expected and caused the bureau to issue a false alarm to the northeastern states. Ione did, however, make landfall again in southern Newfoundland, Canada, and was described as "still packing a powerful punch."

Unlike Connie and Diane, Ione was responsible for several deaths in North Carolina. Five persons drowned across the state, including two in New Bern. One of those was a nine-year-old boy who stepped off his front porch into floodwaters that were over his head. Two additional deaths occurred in mishaps involving automobiles. A Camp Lejeune marine was killed when his car crashed through a barricade set up at a washed-out bridge, and a Beaufort man drowned when his car accidently spilled into a roadside canal that was overfilled by floodwaters.

Most of the damages in North Carolina that were directly attributable to Ione resulted from losses to agriculture. Crops were destroyed, topsoil was washed away, and farmland was contaminated by salt water. It was estimated that almost 90,000 acres of land in the eastern part of the state were submerged during the storm. Approximately $46 million in crop damage was reported, while official estimates of Ione's total price tag in North Carolina approached $90 million.

The three storms of 1955 made that year the most expensive in North Caro-

Ione's winds toppled dozens of large trees in downtown New Bern. (Photos courtesy of the New Bern–Craven County Public Library)

lina's hurricane history. Connie, Diane, and Ione together caused $170 million in property and agricultural damage to the state. When combined with the losses from Hazel the previous year, North Carolina faced an economic hardship of unprecedented proportions. Governor Luther H. Hodges testified before the U.S. Senate Appropriations Committee that the four storms had caused a loss of $326 million. That total exceeded the entire state's annual revenue for the year. As an illustration of this great loss, it was noted that "3000 miles of first-class highways might have been built in eastern North Carolina for the money lost to hurricanes." Relief aid was provided by numerous groups and included federal financial assistance approved by President Eisenhower. Even the Russians helped the flood-ravaged people of the area through contributions from the Russian Red Cross and Red Crescent.

The fact that three hurricanes rolled through North Carolina within six weeks was only one of the distinctions of the 1955 season. Remarkably, during this same period, Tampico, Mexico, was also struck by three hurricanes. In the space of twenty-five days, Gladys, Hilda, and Janet all sliced across the Yucatan Peninsula and onto the beaches near Tampico. Hurricane Janet was a rare category-five cyclone that brought great death and destruction to every location it touched. Janet's winds reached 175–200 mph during portions of its trek across the western Caribbean and Gulf of Mexico. Weather Bureau officials reported that when Janet struck Mexico, "the resulting floods culminated in the greatest natural catastrophe in the history of that country."

The 1955 hurricane season was like no other. Gordon Dunn, chief forecaster for the Weather Bureau in Miami, reported that "for the second con-

secutive year, all records were broken for hurricane destruction." And the storms that rolled through Hurricane Alley were responsible for a large part of the trouble.

HELENE (SEPTEMBER 27, 1958)

Perhaps the most intense tropical storm to threaten the Carolina coast during the fifties was hurricane Helene. Helene passed just twenty miles offshore and barely missed the Cape Fear region on the morning of September 27, 1958. Earlier, the storm had charted a course directly toward the Tar Heel coastline. But fortunately, Helene veered away not a moment too soon, sparing a direct hit near Southport. Also, its approach coincided with low tide, subtracting from the potential flooding it could have delivered.

Even though Helene's most powerful energies remained at sea, new wind records were established at several locations near the coast. At Bluenthal Airport in Wilmington, gusts of 135 mph were recorded, eclipsing the old mark for that station. An observer near Southport estimated winds at 125 mph and gusts to 150–60 mph. Barometric pressures hovered around 27.75 inches, and 8 to 10 inches of rain were dumped along the coast. Helene might have been a more intense storm than Hazel, but because it never made landfall and arrived at low tide, flooding was moderate. Nevertheless, damage to the state was estimated at $11 million.

At Southport, wind damage was significant, and overall destruction was said to have been greater than Hazel in '54. Long Beach suffered as well, when overwash covered roads with sand in some areas and undermined them in others. Nine cottages lost their roofs, and virtually every structure on the island received some damage.

At Wrightsville Beach, a member of the Weather Bureau staff made a careful swell count on the morning the hurricane arrived. Incredibly, Helene produced only two and a half to three giant swells per minute. This report was described as "probably the lowest count ever recorded for the area and indicates a storm of exceptional intensity." Overall, however, tides along the ocean beaches were only three to five feet above normal.

Five thousand residents evacuated their homes from Calabash to Morehead City. Almost all of the reported damages were the result of powerful, gusting winds. Crops were blown down in the fields, and roofs suffered greatly as far north as Cedar Island. Helene would have been even more destructive if its intense northeastern quadrant had moved over the Outer Banks.

Governor Hodges, by this time keenly experienced in hurricane disasters, drove into Wilmington just before Helene passed by. About one hundred residents had refused to be evacuated from Carolina and Kure Beaches, and there was great fear that they might be lost to the storm. Governor Hodges tried to make a television appeal to the remaining beach residents, but he never went on the air. The television station lost power before he could broadcast his plea. Eventually, the governor and his party drove to Wrightsville Beach, where they rode out the storm in a shelter on Johnny Mercer's pier.

No deaths were attributed to Helene in North Carolina, and only a few minor injuries were reported. One woman in Jacksonville was injured when a tree fell through her mobile home, striking her on the head. Few injuries or problems resulted from the storm, because evacuations of vulnerable areas were very complete and cooperation among agencies was well tuned. One official noted, "We were prepared for Helene. You might say that we've had lots of recent experience, with Hazel and Diane, and we knew what to look for."

DONNA (SEPTEMBER 11, 1960)

In the years following the string of hurricane tragedies of the fifties, residents all along the Atlantic coast kept watchful eyes on the storms that brewed during each hurricane season. Storm tracking and forecasting was improving, and hurricane warnings were becoming more reliable. The advent of television helped many residents stay informed about approaching storms. Late in the summer of 1960 the people of eastern North Carolina watched cautiously as another dangerous cyclone raced across the Atlantic and headed toward Florida. This new storm left heavy destruction and scores of dead in its wake

Hurricane Donna struck North Carolina in September 1960, leaving coastal residents with the familiar task of storm cleanup. (Photo courtesy of the News and Observer Publishing Co./ N.C. Division of Archives and History)

Although Donna was one of the most powerful hurricanes of the century when it ripped across Florida as a category four, it arrived in North Carolina as a category three, still potent enough to demolish this block home on Atlantic Beach. (Photo courtesy of Roy Hardee)

even before it reached the United States. It then went on to become one of the most potent hurricanes in history. Its name was Donna.

Early in September, hurricane Donna developed from a tropical wave that moved westward from the Cape Verde Islands off of the African coast. On September 4, it swept over the Leeward Islands and then bounced along the northern coast of Puerto Rico. Before it passed by the Cuban coast on September 7, Donna had already claimed more than 120 lives. As it turned its sights on southern Florida, this monster hurricane packed winds in excess of 150 mph. On September 9, 1960, Donna blasted the Florida Keys with a thirteen-foot storm surge and deadly winds that gusted between 175 and 200 mph. Its barometric pressure dropped to 27.46 inches, making it one of the most intense hurricanes of the century. Several spans of the famous ocean causeway that extended from Key Largo to Key West were washed away by the violent seas. The destruction throughout this region was the worst since the Labor Day Hurricane of 1935.

As Donna crossed into the Gulf of Mexico in the early hours of September 10, its fickle character became evident when it abruptly turned ninety degrees to the north. From that point, the storm slammed into the Florida coast for a second time, striking near Fort Myers, on the Gulf shore. Donna then crossed the state on a northeast track and moved back into the Atlantic near Daytona Beach. Along the way, it brought record destruction to Florida, including heavy damages to the state's citrus crop. In all, thirteen lives were lost in the

Children sleep in the hallway of the Morehead City Town Hall as Donna's eye passes directly over them. Many public facilities were transformed into shelters during hurricane emergencies. (Photo courtesy of Roy Hardee)

The overwash that struck this western section of Atlantic Beach during hurricane Donna left deep sand in the streets and fractured houses all around. (Photo courtesy of Roy Hardee)

Donna left debris throughout the streets of Swansboro.

Sunshine state. But Donna's escapades did not end with its journey through Florida. By the afternoon of September 10, the hurricane was positioned off the Georgia coast, where it regained much of its strength from the warm waters of the Gulf Stream.

Late in the evening of September 11, Donna moved across the North Carolina coastline near Topsail Island. Winds gusted in excess of 100 mph, and tides ran four to eight feet above normal. Wilmington reported a peak gust of 97 mph, but winds at several locations along the Outer Banks were higher.

The eastern overwash at Atlantic Beach struck the Dunes Club head on, ripping the structure apart and carrying portions of its floor and roof across Fort Macon Road. The overwash was later filled in, the roadway restored, and the Dunes Club rebuilt.

Donna continued on a northeasterly course, passing over Carteret, Pamlico, Hyde, and Tyrrell Counties before crossing over Albemarle Sound and through Elizabeth City. Eventually, it returned to the Atlantic for the second time near Virginia Beach on the morning of September 12.

Coastal communities from Carolina Beach to Nags Head suffered extensive structural damage. Beach erosion was severe in some locations, and numerous overwashed areas were reported. Wind damage was severe, especially to crops, as far as fifty miles inland. Numerous trees were toppled, and power outages were typical along the coast.

Carteret County found itself on the eastern side of the storm, and the coastal communities of Atlantic Beach, Morehead City, and Beaufort were among the hardest hit. In Atlantic Beach, several buildings were leveled, including a pavilion and a bakery constructed of concrete block. Numerous structures lost their roofs completely, and many decks and porches were ripped away by high winds. Donna's storm tide cut through the protective dune line in several places, overwashing streets and undermining homes. Breakers ripped through the dunes on Knollwood Drive in Pine Knoll Shores, and a second overwash carried away a twenty-five-foot section of Ocean Ridge

Donna's track northward brought flooding to downtown Manteo. (Photo courtesy of the Outer Banks History Center)

The Dare Hardware Company in Manteo was wrecked during hurricane Donna. (Photo courtesy of the Outer Banks History Center)

Donna destroyed the Waterside Theater near Manteo, home of The Lost Colony. (Photo by Aycock Brown; courtesy of the Outer Banks History Center)

Across eastern North Carolina, countless trees were downed during hurricane Donna. Many fell on power lines, cars, houses, and other structures. This large pine fell into a home near Jacksonville.

Road in Atlantic Beach. But perhaps the most impressive overwash occurred just east of the Oceanna Pier, where the Dunes Club was destroyed. The beach club was broken apart by the waves, and the main floor of the structure was washed onto Fort Macon Road. The roadway was undermined as well, isolating the Coast Guard station and all points east of the overwash.

Some of the heaviest destruction in the Carteret region occurred on the Morehead City–Beaufort Causeway along a low-lying stretch of Highway 70. This area had been hard hit during Hazel, and Donna's winds and tides brought even greater destruction. After the hurricane had passed, reporters flocked to the scene, and one writer for the *Greensboro Daily News* described the aftermath:

> Up and down the causeway almost everything is wreckage. Power and telephone lines are down across the road, the poles snapped off at their bases. The road itself is half-gone in spots, thick chunks of broken asphalt just out of the pits and ravines carved into the sand by the wind-driven tides. The Beaufort and Morehead Railroad runs in a single track along the north side of the causeway, and it is here that Donna did the railroad dirty. The storm cut the sand from under the tracks and left the crossties hanging—barely in some places—by the rail spikes, and the rails themselves dangling over the water like two long strands of half-cooked spaghetti. A diesel engine leans crazily toward the water—still on the tracks, but with nowhere to go. Behind it, on both sides of what used to be the railroad embankment, stretches a bizarre parade of upturned boats, beds, tables, pillows, smashed planking, pilings, beams and driftwood. A pair of water skis, a child's doll, an outboard motor housing, a sofa that got out of its house without, somehow, making a hole in either wall, windows or door.

At the height of the storm, an ambulance was swept off the causeway by the tide. Driver Bert Brooks and his companion, Cecil Moore, were transporting Annie White to the Morehead City hospital. White was expecting a baby and was being taken there for fear that the hurricane might isolate her from the mainland. The rapidly rising tide caught the ambulance before Brooks and his passengers could make their way across the bridge. They emerged from the vehicle in chest-deep water and managed to retreat to a house on the causeway, where they were able to safely ride out the storm. The ambulance was badly damaged and was partially buried in the sand. Annie White didn't have her baby that night after all, but gave birth two weeks later.

The Outer Banks were struck with the full fury of Donna. High winds, estimated to be 120 mph in some locations, ripped away roofs and toppled miles of telephone and power lines. The storm's early southeast winds piled a mass of water up the Pamlico, Albemarle, and Currituck Sounds, inundating their banks and tributary rivers. As the hurricane passed, the winds turned to the

This car was almost flattened by the collapse of a large tree near Kinston. (Photo courtesy of the News and Observer Publishing Co./N.C. Division of Archives and History)

northeast and a raging flood struck the Outer Banks. Dozens of homes on the soundside of Kitty Hawk and Nags Head were severely damaged. At least three houses were swept into Roanoke Sound near the Little Bridge between Nags Head and Manteo. Violet Kellam, owner of the Oasis Restaurant at the Little Bridge, escaped as her business collapsed around her. She told the *Virginian-Pilot* that "the water rose more than three feet in 15 minutes" and the eye lasted "more than an hour." In some areas, it was reported that "telephone poles popped like firecrackers."

Dozens of towns and cities in the east suffered damages from hurricane Donna. In New Bern, many large trees were downed, and some crashed into homes and businesses. In Washington, Edenton, Swan Quarter, and Elizabeth City, fallen trees and toppled power lines were reported, as well as water damage. More than seventy miles of power lines were downed north of Swan Quarter, and most communities in the northeastern counties were without electricity. Donna's ferocious winds also reached inland and struck signs, trees, and telephone poles in Kinston, Goldsboro, and Greenville. Two tornadoes were observed during the storm, one in Bladen County and one in Sampson County.

Like so many other great hurricanes, Donna didn't end its journey by fizzling out over the Atlantic. Instead, it maintained its course toward New England and struck Long Island, New York, later in the day on September 12. There the storm delivered a ten-foot storm surge and caused extensive damage. Block Island, Rhode Island, recorded sustained winds of 95 mph, with gusts up to 130 mph. Damage was reported all along the northeast coast, from Virginia to Maine. Donna eventually weakened when it crossed the Gulf of St. Lawrence on a northerly track.

Hurricane Donna caused a record amount of destruction in Florida and brought significant damage to the storm-weary residents of eastern North Carolina. By hitting New England with a powerful blow, Donna became a menacing oddity—it was the first storm to strike with hurricane-force winds in Florida, the Carolinas, and New England within the seventy-five-year record of the Weather Bureau. Damage estimates in the United States topped $426 million, although Donna's total cost was believed to be more than $1 billion. At least 50 deaths were blamed on the storm in the United States, and 121 lives were lost in the Caribbean.

In North Carolina, there were eight deaths. Three boys drowned when they took refuge in a house that was swept away by the tide. One person was electrocuted by a fallen power line, and two died when they were crushed by large trees that crashed into their homes. Two others were killed in weather-related traffic accidents. Over one hundred injuries were reported across the state.

Donna's arrival in North Carolina just a few short years after Hazel and the other major storms of the fifties was difficult for many coastal residents to take. But very few packed up and moved away from their homes near the shore. In an editorial that appeared in the *Carteret County News-Times* shortly after the storm, the editor summarized the mood of the local people: "Donna left in her wake not only material destruction but crushed spirits. You can fight just so many hurricanes and then the loss of your business, home, plus the drudgery of back-breaking clean-up begins to be a heart-breaking task. Carteret will come back, because there's nothing else to do but that, but the novelty of hurricanes has long worn off."

In the years following the hurricane tragedies of Hazel, Connie, Diane, Ione, and Donna, the coastal waters of North Carolina remained relatively quiet. Several tropical storms and hurricanes passed east of Hatteras, but for more than a decade no substantial hurricane destruction occurred in the state. This mysterious period of calm was a blessing to the many coastal residents who had repaired and rebuilt their homes and businesses after Donna. This period was also one of unprecedented growth along the coast. Numerous beachfront communities prospered as they became widely recognized as major tourist destinations.

THE MODERN ERA,
1960–1999

Track of hurricane Ginger, 1971

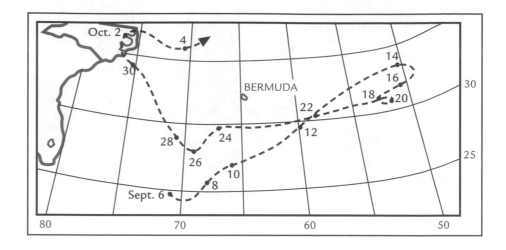

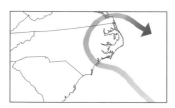

GINGER (SEPTEMBER 30–OCTOBER 1, 1971)

Hurricane Ginger was the first hurricane in several years to test the will of North Carolina's coastal residents. On September 30, 1971, Ginger made landfall on Atlantic Beach as a mild category-one storm. Gusts were reported as high as 92 mph in Atlantic Beach, 70 mph at Cape Hatteras, and 58 mph at Topsail Beach. Tides along the beaches were about four feet above normal, although several locations along the banks of Pamlico Sound recorded tides of five to seven feet.

Rainfall from Ginger was heavy, largely because of the extremely slow movement of the storm. Rainfall totals greater than ten inches were reported at Bayboro, Belhaven, Aurora, and Roanoke Island. The combined effects of heavy rains and wind brought damage to thousands of acres of corn, soybeans, and other crops. The total losses to agriculture alone were estimated at $10 million.

Property damages along the coast were relatively minor. Some trees were uprooted, some power lines downed, and numerous mobile homes overturned. Television aerials, signs, and fences were blown down, and dozens of plate-glass shop windows were shattered. Thousands evacuated oceanfront beaches and low-lying areas and took shelter in schools and other public buildings. Fortunately, Ginger caused no deaths or injuries.

Although Ginger was not as severe as the hurricanes of the fifties, the storm did establish a new weather record—it was the longest-lived hurricane in National Weather Service history. After forming just east of the Bahamas, Ginger moved well to the east of Bermuda and then backtracked toward the Atlantic coast. It was tracked for thirty-one days, during twenty of which it maintained hurricane strength. By the time the storm moved into North Carolina,

134 THE MODERN ERA

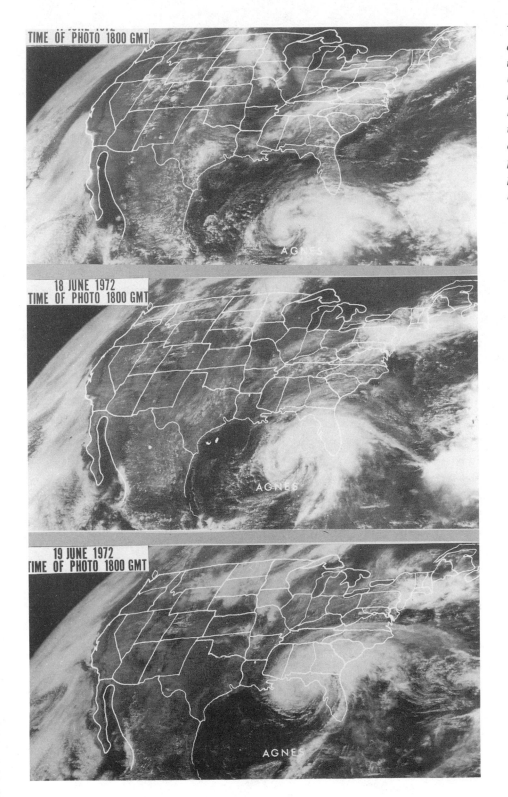

TIME OF PHOTO 1800 GMT

18 JUNE 1972
TIME OF PHOTO 1800 GMT

19 JUNE 1972
TIME OF PHOTO 1800 GMT

AGNES

AGNES

AGNES

The broad, spiralling storm clouds of hurricane Agnes dominate these satellite images of the United States. Although Agnes made landfall near Panama City, Florida, it moved northward through the Carolinas, where it combined with a secondary low from the West to produce record floods. (Photos courtesy of the National Weather Service)

it had dissipated substantially. On October 2, Ginger passed over Virginia and out to sea as a weak depression.

AGNES (JUNE 20–21, 1972)

After Ginger crossed the coast in 1971, there was another relatively quiet period along the North Carolina beaches. Few severe storms threatened the state through the remainder of the 1970s. Most of those that did approach were either weak tropical storms or hurricanes that passed far off the coast.

One exception was hurricane Agnes, which moved through the interior portions of Florida, Georgia, and the Carolinas in June 1972. Agnes did have a severe effect on North Carolina, but it was not the coastal region that suffered. Record rainfall occurred on the eastern slopes of the Blue Ridge Mountains from the Carolinas through Virginia. These heavy rains, which totaled more than ten inches in some locations, brought flash floods to mountain and Piedmont streams and major river flooding across the Tar Heel state. The most severe flooding occurred on the Yadkin–Pee Dee River system and on the Dan River. Extensive flooding also occurred along the Catawba, Saluda, Rock, Congaree, Lumber, and Broad Rivers.

Two deaths were blamed on Agnes in North Carolina. One Iredell County man was swept off his tractor by raging floodwaters, and another life was lost when a canoe overturned in Surry County. Damages in the state were estimated at $4.28 million. But the destruction that occurred from Florida to New York made hurricane Agnes one of the most expensive storms in history. Devastating floods in Virginia, Maryland, West Virginia, Pennsylvania, New Jersey, and New York combined with the destruction of fifteen tornadoes in Florida to raise the price tag for Agnes to well over $2 billion. In all, 122 lives were lost in the United States.

DAVID (SEPTEMBER 5, 1979)

In September 1979, hurricane David swept across North Carolina on a track toward the northeastern states. Like Agnes, David moved through the Piedmont, never actually making landfall on the North Carolina coast. David did, however, bring gusting winds and high tides to the southern beaches, as well as heavy rains to numerous locations.

David's greatest destruction occurred before the storm reached the continent. More than one thousand deaths were blamed on the hurricane as it passed through the Caribbean Sea. Winds of 150 mph lashed Puerto Rico and the Dominican Republic, and heavy rains delivered massive floods to these islands. David's next move was toward the east coast of Florida, where residents in Miami braced for the storm. On Labor Day, September 3, David wobbled erratically, just missed Miami, and struck near Palm Beach. The cyclone then bounced back to sea near Cape Canaveral, Florida, and made landfall again near Savannah, Georgia. From there the hurricane lost much of its energy and passed through central South Carolina and North Carolina as a tropical storm.

Beach erosion was severe along the southern strands of North Carolina, especially in Brunswick County, where most beaches lost thirty to forty feet of sand. At least half a dozen fishing piers were crippled by the storm. Tides were generally three to five feet above normal, although some locations along the banks of Pamlico Sound reported tides of seven feet. Wind gusts of 60 mph were reported at Wrightsville Beach, 54 mph at New River, 53 mph at Atlantic Beach, 43 mph at Cape Hatteras, and 36 mph in Raleigh. Rainfall over several eastern counties ranged from seven to ten inches.

In Goldsboro, tornado-like winds toppled trees in the downtown area. Town employees in Belhaven were forced to evacuate when five inches of water sloshed inside Town Hall. About thirty thousand Carolina Power and Light customers were without electricity when gusting winds pitched tree limbs into power lines. Seven eastern North Carolina counties closed their schools because of the threat of flooded highways. Fortunately, no deaths or injuries were reported in the state.

David continued to deliver heavy rains and gusting winds to the Northeast after the storm passed through the Carolinas on September 6. Virginia, Maryland, Pennsylvania, New York, Connecticut, and Massachusetts were all hit hard by the storm. Numerous tornadoes were spawned on David's northern track, and three twisters caused deaths in Philadelphia, southern New Jersey, and Rhode Island. Throughout its trek of some five thousand miles, hurricane David was responsible for over 1,100 deaths and $2 billion in damages. Like hurricanes Diane in 1955 and Agnes in 1972, David demonstrated the destructive power these storms can possess long after they have tracked inland.

DIANA (SEPTEMBER 9–14, 1984)

Emergency management teams and local officials put their disaster plans into high gear when hurricane Diana approached the North Carolina coast in September 1984. Diana was a fickle storm that finally made landfall near Bald Head Island on September 13 after drifting around Cape Fear for almost two days. Although it had been a minimal category-four storm, Diana was barely a category two by the time it hit the beaches of Brunswick and New Hanover Counties. Nevertheless, it was the first significant hurricane to strike the North Carolina coast since Donna in 1960. The span of time between Donna

2130Z 11 SEPTEMBER 1984

and Diana had been one of the least active in the state's recent hurricane history. As a result, a whole new generation of coastal residents had their first hurricane experience.

Diana began as a small tropical low just north of the Bahama Islands on September 8. By September 10, the low had developed into the first hurricane of the season and was drifting northward about two hundred miles off the Georgia coast. Over the next two days, Diana intensified significantly as it appeared to follow the Gulf Stream toward North Carolina. On September 12, the storm's central pressure reached its lowest value, 28.02 inches. Poised less

than fifty miles off Cape Fear, Diana churned the ocean with sustained winds of 135 mph. At that point it appeared that landfall was imminent and that great destruction would follow.

Hurricane warnings had been issued prior to Diana's approach, and barrier island residents from Myrtle Beach to Ocracoke boarded their homes and evacuated. Thousands took refuge in hotels, schools, and other public shelters. Across the southern coast, residents watched vigilantly as local television newscasters provided around-the-clock updates on the hurricane's movements. But as Diana neared the Cape Fear region, it stalled and wobbled briefly away from land. Some evacuees became impatient after the first night away from their homes and actually returned to the beaches while Diana drifted about. But after some thirty hours, the hesitant hurricane turned and finally made landfall, forcing many to flee for the second time in two days. One Wrightsville Beach resident commented after the storm: "I'm not sure which was worse—the anxiety of worrying about what was going to happen, or the actual storm itself. At least it's over!"

While Diana sat offshore, the once-powerful hurricane lost much of its strength. The highest sustained winds measured on land were 115 mph at the Oak Island Coast Guard station. This report was made on September 11, long before the time of landfall. When the storm did finally move onshore sometime after midnight on September 13, sustained winds were around 92 mph. Gusts throughout the local area surpassed 100 mph.

Fortunately, Diana's stalled movements caused it to weaken, but the timing of its final approach also turned out to be a blessing to beachfront property owners. Landfall occurred very near the time of low tide, and the effects of

Diana toppled many large trees in Brunswick and New Hanover Counties, including this oak in Southport.

Diana's winds buffeted Carolina Beach and caused significant damage to some structures. (Photo courtesy of S. M. Rogers Jr./U.N.C. Sea Grant)

Anne Donnell of Southport searches through the debris in her living room after enduring a frightening visit from hurricane Diana. The storm's gusting winds ripped away a portion of the roof of the Donnell home.

storm surge were minimal. Beach erosion was somewhat severe from the pounding northeast winds, but the storm tide at Carolina Beach was only about five and a half feet. Freshwater flooding was widespread, due to heavy rainfall over a three-day period. Some locations recorded 15.50 inches. The National Weather Service office in Wilmington reported 13.72 inches from September 11 to 14. Dam failures resulted from the torrential rains in Boiling Springs Lake (Brunswick County), Roseboro (Sampson County), and Faison (Duplin County).

The eye of hurricane Diana passed slowly over Southport around 2:00 A.M., after several hours of punishing winds. An eerie calm came over the town as the winds died. While the eye was over Southport it was reported that crickets could be heard chirping in the still night air. Old-timers who could remember the passage of Hazel said there was little comparison; Hazel's eye swept through in only minutes, while Diana's lasted for almost two hours.

Diana weakened quickly after moving onshore and tracked to the northeast as a tropical storm, exiting the coast near Oregon Inlet. Although Brunswick and New Hanover Counties were hardest hit, Pender, Sampson, Bladen, and Columbus Counties also suffered from the storm. A single tornado was reported in Nash County as the remnants of the hurricane returned to sea.

There was widespread tree damage along the coast and in the interior sections near the storm's path. Many large pines and oaks were snapped by gusting winds, but more commonly trees were uprooted and toppled. Heavy rains loosened the soil supporting many trees, aiding their collapse. Downed trees often crashed into homes, blocked streets and highways, and tangled power lines. As a result, some areas were without electricity for days after the storm passed.

Structural damage was widespread but variable. Many newer homes constructed to meet more modern building codes fared well in the storm. Older structures suffered greater damages, including complete roof failures in some cases. In one spectacular incident at Carolina Beach, a large section of roof from a motel flew more than five hundred feet over several buildings, finally crashing into the roof of another two-story structure. Some coastal properties were reportedly "sandblasted enough to damage or remove the paint." In all, the Red Cross estimated that fewer than one hundred homes were completely destroyed, around six hundred were severely damaged, and about "90 percent of the homes touched by the storm received only minor damage."

Agricultural damages from Diana were greatest to corn, soybeans, peanuts, and sweet potatoes. Although total crop losses were near $25 million, the damages were reported to be far less than the crop destruction from the tornadoes that moved through the state in March of the same year. The agricultural damages, when combined with the property losses, brought the total cost of the hurricane to near $85 million. But as one official noted, "You have

Atlantic Beach residents line up at the town hall to be issued their official evacuation reentry passes prior to the arrival of hurricane Gloria in 1985. (Photo courtesy of the Carteret County News-Times*)*

The passage of hurricane Gloria in September 1985 caused significant pier damage along the coast, but overall damage in North Carolina was less than had been expected. The storm caused about $8 million in losses to the state, but well over $1 billion in damages to the Northeast. (Photo courtesy of the Carteret County News-Times*)*

to realize that this could have been far, far worse. This total amounts to small potatoes when compared to other recent hurricanes."

Three deaths were attributed to Diana in North Carolina. One person suffered a fatal heart attack while preparing for the storm, and two died in automobile accidents on water-covered roads. One of those was Brunswick County social services director Larry Bell, who was attempting to make his way to the evacuation shelter he had earlier established.

Diana's impact on North Carolina was not as catastrophic as the impact of the hurricanes of the fifties, and the losses were not nearly as great. Those residents along the southern coast who suffered damages were nevertheless awed by the storm's destructive power. But ironically, this hurricane provided some practical benefits. Coastal communities and state officials were given the opportunity to fully implement their newly developed strategies for mass evacuation and disaster management. In Carteret County, for example, evacuations gave officials a "dry run" that enabled them to make improvements in their plans. And Diana offered the first real wind-effect test of the state's recently modified building codes. Most important, as the first hurricane to strike in many years, Diana delivered a wake-up call to many of North Carolina's coastal residents. They were reminded of the potential for disaster that can be swept ashore with a major storm.

This air-supported phosphate storage building at the state port in Morehead City collapsed from the early morning winds of hurricane Gloria. (Photo courtesy of the Carteret County News-Times)

Motorists leaving the Outer Banks crowd the Wright Memorial Bridge before the arrival of hurricane Gloria. (Photo courtesy of the Carteret County News-Times)

GLORIA (SEPTEMBER 26–27, 1985)

One year after Diana visited the Cape Fear coast, hurricane Gloria sped toward North Carolina and presented another serious threat to the state. Gloria was a larger, more powerful storm, which appeared to be headed for a direct hit near Cape Lookout. Some had even dubbed it "one of the storms of the century." But like many hurricanes, Gloria changed course slightly as it approached the state and crossed the Outer Banks near Cape Hatteras on September 27, 1985.

Gloria was first observed as a tropical depression off the western coast of Africa on September 15. For several days, the storm drifted across the Atlantic and intensified. On September 22, Gloria reached hurricane strength and began a northwesterly course toward the Carolinas. By the time it reached a point four hundred miles southeast of Cape Hatteras, Gloria's central pressure had dropped below twenty-eight inches. For the second time in two years, a category-four hurricane was threatening North Carolina's beaches.

Hurricane warnings were issued, and thousands fled their homes to seek refuge in emergency shelters. Schools, hospitals, churches, and other public facilities were packed with evacuees, just as they had been during Diana. But the warnings for Gloria were more ominous and the evacuations more complete. Even longtime residents of Ocracoke, accustomed to frequent storms and severe hurricanes in the past, left their island homes for safe shelter. Many boarded the ferry to Swan Quarter, where they found refuge in shelters and hotels. It was estimated that less than ninety residents remained on Ocracoke to ride out the storm.

Early predictions called for Gloria to make landfall near Morehead City. But as the hurricane approached the central coast, it swerved northward and passed over Hatteras Island around 1:30 A.M. on September 27. Gloria's eye continued across the banks and reentered the Atlantic near Nags Head on a northeasterly track. From there, the storm regained strength and picked up forward speed. Gloria was moving about 35 mph when it struck the continent for the second time on Long Island, New York. It continued across Long Island Sound and blasted Connecticut, Rhode Island, Massachusetts, and Maine with high winds and heavy rains.

Amazingly, Gloria's sweep of the Outer Banks brought modest damage to the islands. Even though the hurricane had been touted as "the most powerful in years," several factors contributed to the less-than-expected destruction. Fortunately, the lunar tide was receding as the storm brushed over Cape Hatteras. And as with many hurricanes, Gloria's more powerful eastern half remained over the Atlantic as it spun up the coast. Also, Gloria moved through rapidly, never lingering to pile up water and destruction.

Although reconnaissance aircraft had reported winds up to 130 mph, the highest winds measured on land were substantially less. Diamond Shoals Tower, about fifteen miles southeast of Cape Hatteras, recorded maximum sustained winds of 98 mph and gusts to 120 mph. At the Cape Hatteras Weather Station, sustained winds of 74 mph and gusts to 86 mph were measured. Several locations from Ocracoke to Virginia Beach reported gusts over 100 mph. But overall, Gloria's winds were less than had been expected.

Observers at the Cape Hatteras Weather Station recorded a low pressure reading of 27.98 inches, classifying Gloria as a category-three storm at the time of landfall. As the Weather Service staff anxiously watched their radar screens and instruments, the hurricane's eight-mile-wide eye rolled directly over the station. The storm's howling winds dropped quickly to 6 mph, and the calm lasted for about thirty-seven minutes. But when the southern eye wall struck, it hit quickly, and the light winds grew to hurricane force within three minutes. The rapid changes in barometric pressure that occurred with the passing of Gloria's eye caused headaches and "popping" ears in the Weather Service staff.

High tides were most severe along the northern Outer Banks. Highway 12 was overwashed in several locations, and sand covered the roadway near Avon and on the northern end of Ocracoke. Tides were generally six to eight feet above normal on the Outer Banks, six feet in the Cherry Point area, and four feet at Wrightsville Beach. Erosion along the oceanfront was severe in numerous locations, as some beaches lost more than twenty-five feet of dune. Rainfall totals included 7.80 inches at Williamston, 7.09 inches at New Bern, and 7.00 inches at Cherry Point. Although the hurricane tracked directly over the island, Cape Hatteras reported only 2.10 inches.

Dare County suffered the greatest damages from hurricane Gloria. In Manteo, four feet of water flooded downtown streets and several businesses. Two fires blazed out of control on Roanoke Island during the storm. A home on Pond Island, near Manteo, was lost when firefighters were forced to retreat as floodwaters rose. A dangling power line was believed responsible for another fire at the Ace Hardware store on Highway 64. Firemen battling the blaze lost their water supply and were forced to pump water out of nearby flooded streets. By daybreak the store had been leveled.

As with most hurricanes, high winds caused thousands to lose power during the storm. Trees were uprooted, mobile homes were flipped over, and several structures lost their roofs. Numerous fishing piers were damaged by the storm's rolling seas. The double-masted cargo schooner *Jens Juhl* broke from its moorings near Beaufort and snapped its masts as it became wedged under a nearby drawbridge. In Morehead City, high winds ripped open an air-supported phosphate storage building at the state port. Residents were surprised to see the "balloon building" deflated in the aftermath of Gloria.

But the damage in North Carolina from Gloria totaled only about $8 million, far less than expected. In an interview with the *News and Observer* after the storm, state emergency management official Lt. Douglas Hoell Jr. commented on Gloria's toll: "We were amazed at the lack of damage for a storm this size. We're concerned that people are going to look at this hurricane and say this was the fourth worst hurricane ever, and it didn't do any damage. So they may not evacuate next time, and next time may be the killer storm." Only one death was attributed to Gloria in North Carolina, when a tree fell into a mobile home and killed a man.

After Gloria brushed by the Carolina coast, it maintained its course and

As hurricane Charley passes nearby, a Beaufort resident bails out his boat on the waterfront at Taylor's Creek. (Photo courtesy of Scott Taylor)

slammed into the waterfront communities of Long Island, New York. From there the storm passed over Connecticut and sped through central New England, finally dissipating over Canada. Over one million residents were without electricity throughout the region, as 100- to 130-mph gusts ripped down power lines and trees and battered homes. Several tornadoes were spawned as the hurricane tracked over one of the most densely populated regions of the country. At least six deaths were blamed on Gloria in the Northeast, and damages surpassed $1 billion on Long Island alone. Much like the Great Atlantic Hurricane of 1944, Gloria had swept the Outer Banks and then the New England states within just a matter of hours.

Gloria's impact on North Carolina was not significant, but like Diana the year before, it caught the undivided attention of the state's coastal residents. And like so many storms before it, Gloria punished the Outer Banks more severely than other parts of the Carolina coast. Fortunately, its passage left few scars on the Tar Heel state.

Charley's winds ripped open this airport storage facility near Manteo. (Photo courtesy of the Coastland Times)

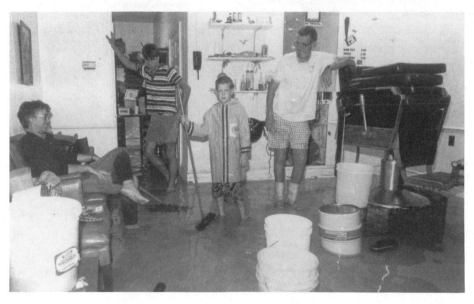

Although hurricane Charley did not cause significant damage to the coast, some areas received minor flooding, including this Dare County barbershop. (Photo courtesy of the Coastland Times)

CHARLEY (AUGUST 17–18, 1986)

Hurricane Charley put North Carolina's emergency evacuation plans to a test when it abruptly crossed the Outer Banks on a busy summer weekend in 1986. Charley originated in the eastern Gulf of Mexico as a tropical depression on August 12. By the fifteenth, the storm had crossed southern Georgia and entered the Atlantic off the coast of Savannah. Then, on August 17, the depression rapidly intensified and threatened the beaches of North Carolina.

Unlike the more powerful Cape Verde–type hurricanes that may be observed for thousands of miles before they reach land, Charley sprang up quickly and caught residents and officials off guard. Very little lead time was available to prepare evacuation shelters or implement emergency plans. One official with the National Weather Service summarized Charley's approach: "It was sitting off the coast of South Carolina as a small low pressure area and then almost explosively, it became a hurricane."

Fortunately, as Charley drifted across Ocracoke, Pamlico Sound, Hyde and Dare Counties, and the Currituck area, it was classified as a "weak" category-one hurricane. The storm exited the state through eastern Virginia before losing its hurricane intensity and returning to sea. Charley was a very short lived hurricane, as it only maintained that status for twenty-four hours. Its 75-mph winds delivered minimal damage throughout the region. Land stations reported sustained wind speeds of less than hurricane force, although gusts to 80 mph were recorded in Hyde County and along the northern Outer Banks. Cape Lookout reported gusts to 58 mph; Wilmington's highest wind was only 29 mph. Charley's greatest winds were reported near the Chesapeake Bay Bridge-Tunnel where a gust of 104 mph was recorded.

Because of Charley's rapid approach, and because much of the storm's intensity was gained overnight, residents and vacationers had very little time to prepare. A voluntary evacuation was announced at around 8:00 A.M. on Sunday, August 17, just hours before the storm skipped across the state. About twenty thousand people left coastal areas within those few hours, causing massive traffic snarls along major evacuation routes in the east. Traffic leaving Atlantic Beach and Emerald Isle was backed up in eight-to-twelve-mile lines from the bridges on either end of Bogue Banks. Farther north, traffic on U.S. 158 was, at one point, frozen in a line twenty-five miles long as anxious travelers fled the northern Outer Banks for the mainland.

The mass exodus left many tourists and residents frustrated. Some sat in traffic for more than three hours and only moved a few blocks. Others gave up on their attempts to leave the barrier islands and returned to their homes and hotels after enduring hours of waiting in traffic. Many feared that they might be trapped in their cars when the hurricane struck. But fortunately, Charley's winds and tides did not pose a significant threat to those who fled the storm.

Charley's overall impact on the state was minimal, with light damage reported to trees and power lines. One death was attributed to the storm—a motorist drowned while attempting to cross a flooded causeway near Manteo. The dollar losses from storm damage were not great, although the loss of revenue from abbreviated vacations may have had a significant impact on the summer tourism trade. But most important, hurricane Charley offered lessons in evacuation planning and demonstrated how quickly tropical cyclones can threaten.

Louise Palmer stands in front of her heavily damaged home in Mayesville, S.C., after the passing of hurricane Hugo in September 1989. Even though the storm's most dramatic effects were on the South Carolina coast, inland residents in North and South Carolina were rocked by high winds. (Photo courtesy of the Charlotte Observer)

HUGO (SEPTEMBER 21–22, 1989)

Few hurricanes of the modern era have caused greater destruction or loss than hurricane Hugo did in September 1989. Hugo was an intense Cape Verde storm that began as a cluster of innocent thunderclouds off the western coast of Africa. On September 10, these thunderstorms became a depression that, while drifting westward, grew into a tropical storm on the eleventh and a hurricane on the thirteenth. Hugo gradually curved to the west-northwest and

In the days before Hugo made landfall, residents along the North Carolina coast made preparations for a possible strike, including visits to building supply stores, which sold plywood for use as window covers. (Photo by Drew Wilson; courtesy of the Outer Banks History Center)

slowed its forward speed as it approached the Leeward Islands, packing sustained winds of 160 mph and a central pressure of only 27.11 inches.

This impressive storm rolled over Guadeloupe on September 17, St. Croix on the morning of the eighteenth, and Puerto Rico on the evening of the same day. Hugo attained category-five status briefly while over the Atlantic, but as it moved through these island nations, it ranked as a four. Throughout the eastern Caribbean, the destruction was massive. Damage estimates for the region exceeded $2.5 billion, and at least forty-one lives were lost.

After battering Puerto Rico with 130-mph winds, Hugo turned to the northwest and developed a course that would ultimately lead to the Carolina coast. By September 21, the cyclone was churning the waters of the Atlantic just a few hundred miles east of Florida. At 5:00 A.M. that day, the storm's maximum sustained winds were 110 mph, but by 5:00 P.M. they had increased to 135 mph. Hugo had increased in intensity from category two to category four in only twelve hours. This killer storm was gaining strength and presented a serious threat to the southeastern coast.

On September 20, a hurricane watch was issued for the beaches from St. Augustine, Florida, to Cape Hatteras. Residents began to prepare for the possibility of evacuation. On the morning of the twenty-first, a hurricane warning was issued from Fernandina Beach, Florida, to Cape Lookout. Later that day, the warning was extended northward to Oregon Inlet. But instead of swinging northward as expected, Hugo slammed into the central coastline of South Carolina near Charleston around midnight, September 21.

Landfall occurred at Sullivan's Island, several miles north of Charleston. From there the storm swept inland, maintaining its northwesterly course. As Hugo passed through the central part of the state it weakened slightly, although winds were still of hurricane force when its eye passed between Columbia and Sumter. By 6:00 A.M. on September 22, Hugo had diminished to tropical-storm strength. At daybreak, it passed into North Carolina just west of Charlotte and carved a path through Hickory and over the Blue Ridge Mountains. The storm continued to accelerate as it passed through the state and was advancing at 40 mph when it moved into extreme western Virginia. From there, the remnants of Hugo continued through West Virginia, eastern Ohio, and western Pennsylvania. The once-powerful cyclone was tracked for two more days as an extratropical storm as it turned across eastern Canada and into the North Atlantic.

Hugo was the most powerful hurricane to strike the United States in twenty years. Not since hurricane Camille in 1969 had a storm of such intensity made landfall in the United States. Hugo blasted the South Carolina coast with sustained winds of over 130 mph. In Charleston, the highest sustained wind was estimated at 138 mph. A ship anchored in the Sampit River five miles west of Georgetown reported a sustained wind speed of 120 mph. One hundred miles inland, gusts of 99 mph were reported at Columbia and 109 mph at Sumter. At Folly Beach, on the weaker, southern side of the storm, sustained winds of 85 mph were reported.

The storm surge on the South Carolina coast was extreme. The highest tide was near 20 feet at Bull's Bay, just north of Charleston. This storm tide was the highest ever recorded on the East Coast. Tides were reported of 16 feet at McClellanville, 13 feet at Myrtle Beach, 12 feet at Folly Beach, and 10.5 feet at Charleston. A fisherman in McClellanville reportedly rode out the storm aboard his shrimp trawler and was said to have "floated over the roofs of two-story houses." Not surprisingly, the impact on waterfront properties was enormous.

Hugo's tremendous power delivered incredible destruction across South Carolina. Extreme tides and high winds knocked bridges off their pilings, stranded yachts in the middle of highways, and toppled television broadcast towers. The 150-mile-wide swath of the storm was especially devastating to the forests of the state, as more than six billion board feet of timber were destroyed. That total was more than three times the loss experienced with the Mount St. Helens volcanic eruption in 1980. The Francis Marion National Forest just north of Charleston was among the hardest-hit areas. The U.S. Forest Service estimated that Hugo damaged or destroyed 70 percent of the trees in the 250,000-acre forest, or about one billion board feet of timber. Very

Massive oaks and other hardwoods were toppled throughout the Charlotte area during hurricane Hugo. Cleanup of the downed trees took many weeks. (Photo courtesy of the Charlotte Observer)

Following hurricane Hugo, children stare in amazement at the fallen trees on a neighbor's home in Charlotte. (Photo courtesy of the Charlotte Observer)

few of the uprooted and splintered trees were harvestable, and the economic losses of the timber alone were over $1 billion.

In North Carolina, Hugo had a severe impact, both on the beaches of Brunswick County and in the cities and towns in the western portions of the state. Damage was reported in twenty-nine counties, most of which were designated as federal disaster areas. As the center of the storm rolled past Charlotte, wind gusts of over 85 mph buffeted the region. Trees crashed into homes, cars, and power lines, and utility poles snapped. Charlotte lost more than eighty thousand trees to the storm, many of which were more than seventy years old. Ninety-eight percent of the city's residents lost power, and for some, repairs were not made for more than two weeks. Power outages caused large amounts of raw sewage to bypass treatment plants and flow into streams throughout Mecklenburg County. North Carolina's largest metropolitan area was brought to its knees by the storm.

Numerous other cities and towns felt Hugo's wrath as it crossed the state. Gastonia, Monroe, Lincolnton, and Hickory were all hit hard by the storm. Two to four inches of rain fell across the western counties, although Boone received almost seven inches. High winds ripped down power lines throughout the region, and forests in some areas were leveled. In North Carolina, Hugo damaged more than 2.7 million acres of forests in twenty-six counties, with almost complete destruction of 68,000 acres. Timber losses to the state were valued at $250 million. And like South Carolina, very little timber was salvaged. Forestry experts were overwhelmed by the sheer volume of dead trees.

Hugo's effect on a Charlotte
home. (Photo by Jane Faircloth;
courtesy of Transparencies, Inc.)

Most of the timber was either splintered by the storm or decayed before loggers could reach it.

In the wake of the storm, Tar Heel residents emerged from their homes in awe of the destruction. So many trees, tree limbs, and utility poles were downed that they completely filled the streets and yards of some neighborhoods. Cleanup efforts began almost immediately but were slowed one week after Hugo when seven more inches of rain fell across several western counties. Chainsaws were essential in clearing streets and lawns, but perhaps the most valued commodity in the aftermath of the storm was one we often take for granted—electricity.

From the South Carolina coast to the hollows of the Blue Ridge Mountains, over 1.5 million people were without electric power the morning after the storm. Utility companies scrambled to put crews to work repairing and replacing downed poles and lines. In some areas, the destruction was so extensive that it was difficult to tell where the lines used to be. Line crews, equipment, and supplies were brought in from around the nation to assist with the effort. Some areas were back in service within days, but others were without power for weeks. Many crews worked sixteen-hour days and spent

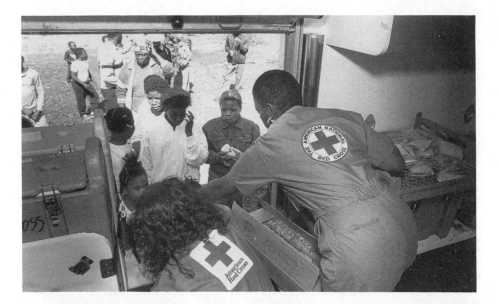

Relief efforts after hurricane Hugo extended through portions of western North Carolina that had been affected by the storm. (Photo courtesy of the Charlotte Observer)

Dare County residents gather food and supplies for shipment to the storm-ravaged victims of hurricane Hugo. (Photo courtesy of the Coastland Times)

weeks away from their families to get the job done. Most residents were appreciative of the line crews' heroic efforts, but tempers sometimes flared as some customers remained in the dark while their neighbors' lights were on.

Duke Power Company was just one utility that reckoned with Hugo's aftermath. In the effort to restore electricity to 700,000 of its customers in North and South Carolina, Duke spent $62.5 million in just two weeks. More than 9,000 workers replaced 8,800 poles, 700 miles of cable and wire, 6,300 transformers, and 1,700 electric meters. The overall recovery effort was unprecedented in North Carolina history. The importance of restoring electricity after any hurricane was highlighted in a government report following Hugo: "In any large-scale natural disaster, energy is the common denominator. Its loss is capable of causing severe economic dislocation. On the other hand, it is essential to recovery as well. In the case of hurricane Hugo, electric power was the principal infrastructure component that had to be rapidly restored. Because the prolonged disruption of electric power can have profound adverse effects on health, safety, commerce and industry, emergency planners must be prepared to respond accordingly."

The southern beaches of the North Carolina coast also suffered from the effects of Hugo. In Brunswick County, the storm produced an eight-to-ten-foot storm surge that battered beachfront cottages. Over 120 homes on Holden Beach, Ocean Isle Beach, and Long Beach were either destroyed or condemned because they suffered extensive damage. Severe beach erosion affected these barrier islands, in many cases washing away the protective dunes that lined their shores. Oceanfront fishing piers were bashed and damaged in Brunswick, New Hanover, Pender, and Onslow Counties. But the most significant of the state's coastal destruction occurred in Brunswick County, where damage estimates topped $75 million.

At the time Hugo struck in 1989, it was the most expensive hurricane in U.S. history. Approximately $10 billion in property was destroyed, $7 billion on the U.S. mainland, $2 billion in Puerto Rico, and $1 billion elsewhere. In North Carolina, the price tag was around $1 billion, even though the storm made landfall in Charleston, some two hundred miles away. As a result, Hugo became North Carolina's most expensive hurricane to that date, although hurricane Fran shattered Hugo's mark in 1996.

The total number of deaths associated with the storm was estimated at eighty-two, including seven in North Carolina, twenty-seven in South Carolina, six in Virginia, and one in New York. Several lives were lost during the cleanup efforts, which lasted for months after the storm. Dozens of injuries were reported ranging from severe cuts and broken limbs to heart attacks. Cleanup crews often encountered live power cables, and chainsaw-related injuries were common. But considering the widespread destruction Hugo brought to the Carolinas, the death toll could have been much higher.

Hurricane Hugo was one of the greatest natural disasters to ever affect the United States. Like Hazel, Camille, and a handful of other hurricanes, Hugo didn't lose its punch when it struck the coast but instead barreled inland with almost full fury. The people of Mecklenburg County thought they were immune to hurricanes prior to this storm's arrival. Most had believed that tropical cyclones were strictly a coastal phenomenon, but Hugo proved to be an exception. After breaking all dollar records for destruction, this incredible storm was overshadowed just three years later by another record-breaking hurricane that more than doubled Hugo's toll: hurricane Andrew.

EMILY (AUGUST 31, 1993)

When hurricane Emily staggered toward the North Carolina coast in late August 1993, tourists and residents along the Outer Banks gave careful thought to the call for evacuation. Along with the rest of the nation, they remembered the vivid scenes of destruction left in the wake of two recent national tragedies. Hurricanes Hugo in 1989 and Andrew in 1992 had generated a healthy new respect for these storms, and many vacationers were not about to take any chances. On August 29, 1993, a voluntary evacuation was announced along much of the North Carolina coast. The next day, the National Hurricane Center in Coral Gables, Florida, issued a statement that included a probability forecast: The chance of Emily making landfall between Wilmington and Norfolk was 21 percent or greater.

On Hatteras Island and Ocracoke, officials made a tough call—they ordered a mandatory evacuation, even though Emily was still almost thirty-six hours away. In years past, some evacuation decisions had been decidedly unpopular, causing merchants to lose tourist dollars and vacationers to lose their time on the beach. Resentment often followed when hurricanes changed course and evacuations turned into false alarms.

But as Emily tracked steadily toward the banks, a stream of cars flowed along Highway 12 and off Hatteras Island. On Ocracoke ferries were filled to capacity, carrying tourists and residents to the mainland. The State Highway Patrol estimated that as many as 90 percent of the people on the banks complied with the call for evacuation. Approximately 120,000 people made the move to avoid the storm. It was estimated that only about 1,000 remained on Hatteras Island to greet Emily when it arrived.

When hurricane Bob slashed past the Outer Banks in 1991, some estimates placed the evacuation response at only 50 percent. Officials conceded that the horrifying images of South Florida in the aftermath of Andrew played a role in the increased concern. "There's no doubt about it," declared one Buxton resident. "Hugo and Andrew showed us how bad these things can be. I have no desire to stick around and find out for myself."

Hurricane Emily did eventually brush Hatteras Island around 7:00 P.M. on August 31, striking the area with sustained winds of 92 mph. The storm was by no means comparable to Andrew or Hugo but was instead rated as a minimum category three, with maximum sustained winds of 115 mph. Fortunately, the higher winds remained just offshore, as Emily's thirty-mile-wide eye came within thirteen miles of Cape Hatteras. Nevertheless, the villages of Hatteras, Frisco, Buxton, and Avon were battered by the storm as it bumped northward along the Outer Banks. On the following day, September 1, Emily turned eastward off the Virginia coast and tracked into the cooler waters of the North Atlantic.

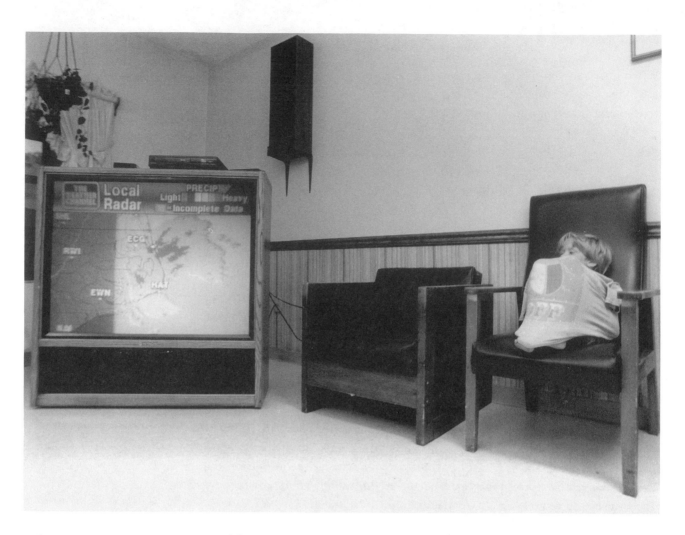

As hurricane Emily churns over the Outer Banks, a frightened young resident of Nags Head studies a radar update. (Photo courtesy of Drew Wilson/Virginian-Pilot/ Carolina Coast)

Although the eye of the storm never actually touched land, Emily's impact on the people and property of the Cape Hatteras area was traumatic. Gusting winds, unofficially clocked at more than 111 mph, snapped ninety-foot pine trees, toppled small buildings, and ripped away roofs. Six Coast Guard family homes in Buxton were leveled by what was at first believed to be a tornado. Later analysis showed that straight, sustained winds had caused the destruction. Mobile homes were rolled over by the blasts of wind, and others had their roofs peeled back "like cans of sardines." One resident watched as a telephone booth rolled off its foundation and into the street.

Most of the major wind damage occurred to older homes or to buildings that were not built to code. The most common structures affected were signboards, porches, poorly connected roofs, eaves, storage sheds, and mobile homes. If Emily had not been deflected by an approaching cold front, landfall might have brought much higher winds and significantly more destruction.

Emily's winds ripped across Pamlico Sound and piled waist-deep water in

The strong winds and high tides that battered the Outer Banks during hurricane Emily were the most severe to strike the region in many years. (Photo courtesy of Drew Wilson/Virginian-Pilot/ Carolina Coast)

As with many hurricanes affecting the Outer Banks, Emily's most severe floods came from the rising waters of Pamlico Sound. (Photo courtesy of Drew Wilson/ Virginian-Pilot/Carolina Coast)

Campers scattered by hurricane Emily near Frisco, N.C. (Photo courtesy of Drew Wilson/ Virginian-Pilot/Carolina Coast)

the streets and homes of several Hatteras Island villages. The flooding was made worse by the occurrence of a full moon, which brought even higher tides to the region. Cars were "floating" in several parking lots in Buxton. In numerous homes, waist-high waves broke through windows and rolled into living rooms. The flooding was about one to two feet higher than predicted on one-hundred-year flood maps, resulting in the need for revised maps for Hatteras Island.

By 7:00 P.M. on August 31, the Dare County Emergency Operations center in Buxton had to be abandoned because of the rising tide. Four feet of water flooded the building, forcing county and local officials to retreat to higher ground. Cape Hatteras School was hard hit by the tide, which filled the hallways and classrooms to a depth of four feet. Among the losses were computers, copiers, and thousands of brand-new textbooks. Total damages at the school alone exceeded $3 million.

Flooding along the shores of Pamlico Sound was about ten and a half feet above normal from just north of Buxton to Avon. At Frisco and Hatteras village, the tide was about eight and a half feet. Along portions of Highway 12, sound waters came within one vertical foot of breaching the oceanfront dunes. Surprisingly, the oceanside storm surge was moderate, breaking through the dunes in two locations south of Frisco.

In the wake of the hurricane, residents of the storm-ravaged areas began the arduous task of cleaning up their homes and businesses. About seven hun-

A crew of Coast Guard workers makes temporary repairs to a house near Buxton after hurricane Emily. (Photo courtesy of Drew Wilson/Virginian-Pilot/ Carolina Coast)

Store owner Ray Couch looks on as workers begin cleanup efforts near Cape Hatteras after hurricane Emily. (Photo courtesy of Drew Wilson/Virginian-Pilot/Carolina Coast)

Emily's destruction was focused on the Outer Banks, where $12.5 million in damages was reported. (Photo courtesy of Drew Wilson/ Virginian-Pilot/Carolina Coast)

dred buildings were badly damaged or destroyed, leaving at least 25 percent of the villages' residents homeless. Damage estimates were near $13 million for the seventeen-mile stretch hardest hit by the storm. Cleanup would take weeks, and in some cases months, as neighbor helped neighbor to rebuild and recover.

As with most every hurricane disaster, volunteers and relief workers poured in to assist in the recovery effort. The American Red Cross, the Salvation Army, the National Guard, and the Southern Baptist Convention offered water, food, shelter, and other assistance. Dozens of state forestry workers removed downed trees and limbs and hauled them away. Utility crews worked to repair electrical lines, broken water pipes, and other services on the island. These agencies and others were quick to respond to the needs of the victims of Emily.

But hundreds of other volunteers also pitched in during the days and weeks following the storm. More than two hundred teachers from Manteo and the northern beaches climbed aboard buses and came to the aid of the devastated school in Buxton. With mops and brooms they scoured the classrooms and offices of the school, enabling classes to begin just two weeks after the flood. Schools across the state held fund-raisers to help with the replacement of equipment and textbooks.

Contractors from around Dare County donated lumber, shingles, and work crews to patch and repair damaged homes and businesses. Church groups arrived the day after the storm with supplies and equipment to help clean and

A determined fisherman tests the wave-tossed waters behind the Showboat Motel in Atlantic Beach, despite the advances of hurricane Gordon in November 1994. Gordon eventually turned away from the Carolina coast. (Photo by Lisa Taylor, courtesy of the Carteret County News-Times*)*

Hurricane Emily approaches the North Carolina coast on August 31, 1993. (Photo courtesy of the EROS Data Center)

dry out flood-soaked homes. Some residents whose cars were lost to the storm were given loaners by friends on the mainland. Those who suffered most were moved by the generosity of the people who came to their aid. "God bless all of you," exclaimed one elderly Avon resident as workers cleaned her home. "I don't know how we could have managed without your help. Emily may have hurt us but we still have our health and our friends. We're very lucky."

Although no lives were lost on Hatteras Island, Emily delivered the hardest blow to the banks of any hurricane in over thirty years. Businesses were devastated by the storm. Some got back on their feet quickly and displayed crudely painted "open" signs on the plywood that had once boarded their windows. Other businesses took longer to recover but were open in time for the busy fall fishing season. One popular sales item all along the Outer Banks was the now standard T-shirt proclaiming "I Survived Hurricane Emily." Those who actually earned it by enduring the storm wore the message with pride.

BERTHA (JULY 12–13, 1996)

Following close on the heels of the near-record-breaking 1995 hurricane season and the state's close call from hurricane Felix, 1996 quickly established itself as a benchmark year for deadly weather in North Carolina. Not only was the state paralyzed by winter storms that caused twelve fatalities from January through March, but a record fifty-one tornadoes touched down during the year, many of them associated with powerful thunderstorms in April and June. But by far, the biggest weather stories of 1996 were hurricanes Bertha and Fran. For the first time in forty-one years, two hurricanes struck the Tar Heel coast during the same season, in the process changing our collective perceptions about North Carolina's vulnerability to hurricanes.

The first of the year's direct hits was hurricane Bertha. As families and friends gathered for Fourth of July celebrations along the Carolina beaches, a tropical wave drifted off the African coast and began to churn across the Atlantic. Soon after the last fireworks displays were over, the National Hurricane Center quietly issued its first advisory on the second tropical depression of the year. On the following day, July 5, the depression strengthened and became tropical storm Bertha. Bertha was still more than a thousand miles east of the Lesser Antilles and was virtually ignored by those who continued to enjoy the long holiday weekend.

By 8:00 P.M. on July 7, Bertha had crossed the open Atlantic and was poised just east of Antigua, threatening to strike the Leewards with hurricane-force winds. The center of hurricane Bertha passed between Antigua and Barbuda on July 8, crossing St. Barthelemy, Anguilla, and St. Martin, just north of St. Thomas. Winds at St. Maarten were clocked at 81 mph, but the storm was strengthening. After passing north of Puerto Rico with winds of 90 mph, Bertha gradually turned toward the northwest, slowed its forward speed, then strengthened to category-three status. Reconnaissance aircraft reported sustained winds of 115 mph on July 9. Hurricanes don't often build to this strength in early July. The last storm to do so this early in the season was hurricane Alma in 1966.

Bertha continued its gradual northward turn over the next two days, with its track becoming north-northwest. Its forward speed slowed considerably, and it lost some of its intensity as it paralleled the Florida coast about 175 miles offshore. At 5:00 P.M. on July 10, the National Hurricane Center placed the mid-Atlantic coast under a hurricane warning. The next day, Bertha was 300 miles south of Wilmington, with sustained winds of 80 mph, moving north at about 10 mph. Evacuations were already underway on the Outer Banks, and thousands of vacationers heeded the warnings and fled the state's other barrier beaches.

By late afternoon on July 11, towns like Wrightsville Beach and Emerald Isle

were eerily quiet, except for the idle of police cruisers that roamed the near-empty neighborhoods in search of the uninformed. In communities across the eastern part of the state, about forty evacuation shelters were established in nineteen counties, ready to serve those who had no other place to turn. The turnout was not as great as expected in some shelters, however. Many tourists left their beachfront resorts for the privacy and comfort of hotel rooms located along the major evacuation routes. Even before Bertha's arrival, its economic impact was under way—some businesses prospered, while others suffered as vacationers scrambled to leave the coast. But some tourists and many coastal residents either stayed put or didn't go far—and they went to bed on the night of July 11 with little fear of the approaching category-one storm.

Then, as has happened many times before, the lumbering storm rapidly intensified overnight, due at least in part to its proximity to the Gulf Stream. At 2:00 A.M. on July 12, Bertha was 215 miles south of Wilmington, with winds still measured at 80 mph. By 8:00 A.M., winds were up to 90 mph, and by 1:00 P.M., they had increased to 105 mph. Bertha's sustained winds had intensified by 25 mph in less than twelve hours. The storm was now a category two, and it picked up forward speed just before it slammed into the coast between Wrightsville Beach and Topsail Island at about 5:00 P.M. Bertha was the first July hurricane to strike North Carolina since 1908.

After making landfall, Bertha quickly diminished in strength and fell below hurricane status as it moved inland across eastern North Carolina. It tracked north into eastern Virginia and slid along the northeast coast to New England. Although it had been downgraded to a tropical storm, winds approaching hurricane strength were reported in nearby Atlantic waters from North Carolina to New England. It was declared extratropical on July 14, when the storm center moved from the coast of Maine to New Brunswick, Canada. From there, it was tracked for three more days over cooler waters near Greenland.

On the North Carolina coast, Bertha provided thousands of residents and vacationers with their first real taste of hurricane wind and water. According to the National Hurricane Center's report on the storm, a reconnaissance aircraft's flight-level wind recording of 125 mph, measured several hours before Bertha made land, was the basis for estimating sustained surface winds of approximately 105 mph on the coast at landfall, when the minimum barometric pressure was 28.73 inches. As is often the case with landfalling hurricanes, few reliable measurements were available to verify the estimated winds. Although there was an unconfirmed report of a 144 mph gust at Topsail Beach, most verified measurements were well below 100 mph. Frying Pan Light Station, off the coast of Cape Fear, encountered the eye wall and recorded sustained winds of 83 mph and gusts to 108 mph. The highest winds on land were measured at New River near Jacksonville, where sustained winds of 81 mph and a gust of 108 mph were recorded at 4:21 P.M. Gusts of 100 mph were recorded at Broad Creek in Carteret County, 89 mph at Casper's Marina in Swansboro, 87 mph by Greenville Utilities, and 77 mph at the Duke Marine Laboratory on Piver's Island near Beaufort.

Bertha's gusting winds whipped trees, signs, shingles, and loose objects along much of the central coast. In many communities, the winds were the strongest experienced in decades. Both man-made structures and natural vegetation were put to the test. Particularly hard hit were the communities that felt the storm's eastern eye wall, especially in Onslow, Carteret, Jones, Craven, Pitt, Beaufort, Martin, Bertie, and Hertford Counties.

Heavy rains pounded the Carolina coast as Bertha moved inland and followed the storm as it tracked toward the Canadian Maritimes. Numerous rainfall measurements exceeding 8 inches were reported, including an account of more than 14 inches from near Hofmann Forest. Other totals from the storm's passage over North Carolina included 6.5 inches at Broad Creek, 5.72 inches at Snow Hill, 5.55 inches at Washington, 4.88 inches at Kinston, and 4.86 inches at Greenville.

Bertha's winds and rains were significant, but worse were the storm's surging tides, which caused extensive beach erosion and structural damage for miles around the area of landfall. Hardest hit were the beaches and sounds between Cape Fear and Cape Lookout, especially those north of Wrightsville Beach. Fishing piers were damaged or destroyed from Kure Beach to Carteret County, and miles of oceanfront dunes were stripped of their sand. The highest storm surge, estimated to be between five and eight feet, struck in Pender and Onslow Counties and caused extensive structural damages at Topsail Beach, Surf City, North Topsail Beach, and Swansboro. The surging tide also overfilled sounds, creeks, and inland rivers, destroying hundreds of residen-

tial piers and boat docks. Nowhere, however, was the storm's destruction more evident than in the precariously low-lying resort community of North Topsail Beach.

Bertha's thirty-five-mile-wide eye moved inland just below Topsail Island, focusing the storm's worst effects on the narrow barrier island. North Topsail Beach, viewed by some as having too little elevation for residential development, was hardest hit by the ocean surge. Waves rolled over the area's modest dunes and flooded S.R. 1568, washing away tons of sand and causing the road to collapse in at least three sections. As the storm passed, pools of water up to ten feet deep crossed the broken highway that provided the only access to the northern tip of the island. More than two miles of the road's northernmost end were buried in sand. According to the *Raleigh News and Observer*, about fifteen families were stranded in their cottages beyond the scuttled road, and emergency personnel considered sending a Marine Corps helicopter to rescue them.

One couple from Ohio barely escaped a confrontation with the tides. After deciding to ride out the hurricane in their North Topsail home, they apparently changed their minds as the storm reached its peak. They piled into their minivan and attempted the dangerous trek toward higher ground. By that time, however, the water had risen too high, and they were forced to abandon their swamped-out vehicle and wade back to their home through chest-deep water. The following day, rescue crews pulled their minivan out of several feet of water and sand.

Bertha's surging tide undermined the main road through North Topsail Beach. (Photo by Don Bryan; courtesy of the Jacksonville Daily News)

As Bertha bore down on Topsail Island, police in Surf City rescued more than fifty people who had belatedly decided not to weather the storm. According to a report in *The Savage Season*, published by the *Wilmington Star-News*, Police Chief David Jones endured another of the storm's close calls. "We went to get a lady that was up at the north end, and we got out of the car trying to get to her house, and the roof picked up off her house and just missed the patrol car and us by about 6 feet," Jones reported. "A whole roof crashed in the road in front of us."

North Topsail lost more than just sand dunes and s.r. 1568. More than 120 homes were destroyed, and hundreds more suffered damage from wind and water. Some structures remained visibly intact, even though their ground-floor garages and utility rooms were swept clean by the rush of water. Dozens of washers, dryers, heat pumps, and other appliances were washed into ditches and marshes, forming instant landfills. Scattered among the broken and exposed water mains were cars, buried to their door handles in deep sand. The walls of some homes were peeled open, their contents visible to the outside like oversized doll houses. These were the scenes that attracted television news crews in the days and weeks following Bertha, as North Topsail Beach was generally considered to have borne the brunt of the storm's fury.

Several nearby communities suffered similar destruction, including Surf City and Topsail Beach. Fishing piers were heavily damaged, and severe erosion swallowed up much of the beach. Among those who weathered the storm at Topsail Beach was syndicated television talk show host Rolonda Watts, who disregarded the order to evacuate and remained in her vacation home. Watts, a North Carolina native, stayed put as the storm surge ran underneath her home, which is erected on stilts. "I love this stuff. There's nothing greater than being in the mix," she told the *Jacksonville Daily News*. In the hours before Bertha approached the Carolina coast, she wanted to bring her satellite trucks down to the area to tape an episode for her *Rolonda* show, but the impending evacuation made that difficult. Fortunately, the television personality and her home were both unharmed by the hurricane.

Elsewhere in Onslow County, gusting winds ripped away the roof of the Dixon Middle School gymnasium, which happened to be serving as an emergency shelter for local evacuees. This caused some tense moments for the frightened refugees, who were quickly moved to the school's cafeteria. In other communities such as Hampstead, Jacksonville, and Richlands, pines cracked and oaks collapsed across countless homes, cars, and roadways. Bertha's gusts claimed a landmark pecan tree on Main Street in Swansboro, along with more than one hundred Bradford pear trees in Jacksonville that had been planted as memorials to those who have died in military service.

In Swansboro, a storm surge of six to eight feet flooded much of the water-front area. The White Oak River, which reportedly had waves of eight to ten feet, came just shy of reaching Elm Street, which is normally two blocks away from the river. Among the flooded structures were the Crab House and Snap Dragon restaurants. Upriver, virtually every private dock was torn to splinters and tossed along the river banks in unrecognizable forms. This region saw some of the storm's worst inland flooding.

Farther south, damages were widespread in Brunswick, New Hanover, and

Pender Counties. Along the Brunswick County beaches, from which close to 100,000 people were evacuated, damages were light. In Southport, a portion of the roof of Town Hall was reportedly "blown away." Residents there were also sorrowed to lose the cedar tree at the Whittler's Bench, a local landmark known as the spot where fishermen used to rest, whittle, and gossip. According to Leila Pigott, longtime resident and charter member of the Southport Garden Club, the cedar was planted in 1954, shortly after hurricane Hazel tore down two poplar trees that grew there. But most residents in Southport felt lucky after Bertha, because they managed to dodge the stronger eastern side of the storm.

On the other side of the cape, Kure, Carolina, and Wrightsville Beaches did not fare as well. Fishing piers throughout the region were heavily damaged, and numerous homes and businesses lost their roofs. At Carolina Beach, parts of Carolina Avenue were under three feet of sand. A sewage pumping station failed, and raw sewage flowed into a street near Boardwalk. Much of the recently replenished berm was washed away. Among the notable losses was the Ferris wheel at Jubilee Park amusement center. The forty-two-year-old wheel toppled over during the storm, crashing into a nearby merry-go-round and landing in a pile of twisted metal. The park's owners, Larry and Ginny Spencer, had no insurance on the wheel but managed to maintain their sense of humor about the loss. Soon after the hurricane had passed, a sign was erected near the park's main gate that read "used Ferris wheel for sale."

Like numerous other storm-affected island communities, Carolina Beach was the scene of some bitter disputes on the morning after the hurricane. Tempers flared when town officials refused to allow eager property owners to return to their homes on the island. Soon after dawn on July 13, a standoff took place at the Snow's Cut bridge when residents learned they would not be allowed back until the streets were deemed safe. At the heart of the controversy was Carolina Beach town manager George Rose, who told the *Wilmington Star-News*: "Are we supposed to let people on the island with nails and raw sewage in the streets?" Some frustrated residents posed as work crews and town employees to sneak across the bridge, while others took a more vigorous approach—they parked their cars and swam across the Intracoastal Waterway to check on their families' homes. After an all-day debate, the town reluctantly allowed residents to return at 6:00 P.M.

At Topsail Beach, a similar war of words was under way the morning following the storm. But damages were extensive, and town officials knew they couldn't just open up the island for residents to survey on their own. A 6:00 P.M. curfew was imposed, and few residents were even allowed on the island until Monday, July 15. When town officials finally agreed the area was safe enough to visit, hundreds turned out. They weren't allowed to drive themselves over the island, though—they had to take buses provided by the Ons-

low County School District. Residents who took the "disaster tour" were glad to finally survey their homes but frustrated that they couldn't jump off the bus and get to work.

In Wilmington, most of the damage came with the crack and thunderous collapse of large trees that succumbed to Bertha's winds. Many stately oaks and other hardwoods were lost in the historic downtown district, and around the city, power lines were toppled by falling branches and limbs. One of the notable trees affected by the storm was a large oak on Fifth Avenue, in front of the home owned by Bob Lane and Bob Warren. The tree was perhaps two hundred years old, and Lane even speculated that it might have been one of the original "boundary oaks" used to identify the boundaries of old Wilmington. Before the storm, the massive tree spread well over Fifth Avenue, providing shade to much of the neighborhood. Bertha's gusting winds split the tree down the middle, sending half of it into the street and pulling a water main up out of the ground. More surprising than the damage, however, was what spilled out of the belly of the oak as it crashed to the pavement. Hundreds of pounds of bricks, cobblestones, and Civil War–era cannonball fragments emerged from a hollow in the tree's trunk. Lane said they had been put there around the turn of the century to help stabilize the tree. Placing ballast in a tree's hollow was once common practice, but as this oak grew, its contents were pushed upward. When winds from the storm peaked in Wilmington, the weight of the stone and iron helped to pull the tree over. Lane also noted that the cannonball pieces were mortar shell fragments collected from the dunes surrounding Fort Fisher.

Reports of significant destruction and minor injuries were scattered among many cities along Bertha's course, especially those near rivers and sounds. Marinas and docks were battered in New Bern, Oriental, Washington, and Belhaven. Waters from the Pungo River flowed into Belhaven's streets, flooding homes and businesses and leaving behind a mud line on the walls of the old town hall. In New Bern, Elizabeth Civils witnessed the violence of the storm from her historic home on the Neuse riverfront. Civils, 88, couldn't hear well but said that she could feel the power of the hurricane. During the height of the storm, high winds blew a tree through her bedroom window, scattering broken glass and soaking her furnishings. "The house did all right," Civils told the *News and Observer*, "until this — this thing — came up in the front yard and started banging on the house. It made the whole house shake. I could feel it." What came banging at her door was a thirty-foot Catalina sailboat that had been moored behind the nearby Comfort Suites hotel. The boat, along with a large portion of a floating concrete dock, came crashing over Civils's seawall, knocked over a large pecan tree, and bashed into her home.

Damages were also reported in Emerald Isle, Atlantic Beach, and Morehead City, although the destruction lessened east of Beaufort. Most of the losses were from downed trees, toppled gas station canopies, flattened signs, and damaged docks and piers. Ocracoke and Hatteras Islands were spared the worst of the storm, and residents and vacationers there were allowed to return the day after the hurricane.

In Washington, the dinner boat *Pamlico Queen* broke from its moorings and crashed into the U.S. 17 bridge over the Pamlico River. Wilmington residents gathered in the darkened lobby of the Wilmington Hilton upon hearing the news that a large U.S. Navy destroyer had broken from its moorings in the Cape Fear River and might drift upriver and damage the U.S. 74/76 bridge. Coast Guard crews worked heroically in the driving wind and rain to attach new lines to the ship. A spokesman said later that the bridge was never in any real danger.

Away from the more visible destruction in beach and waterfront communities, inland counties suffered extensive crop damages. Bertha arrived at a time when both corn and tobacco were vulnerable — corn was still green but otherwise mature, and tobacco was ready for harvest. Strong winds blew down acres of plants in the fields, snapping corn stalks and tearing away valuable tobacco leaves. Even tobacco that had already been harvested was lost when power outages caused packed leaves in curing barns to go sour.

Overall rainfall from Bertha was actually less in many areas than the National Weather Service had anticipated, which proved to be a blessing to many. Environmental concerns over potential hog lagoon spills eased somewhat after the storm, even though there was at least one large spill in northern Craven County. Heavy rains and strong winds pushed the lagoon at Rhodes Livestock Farm through a clay dike, spilling 1.8 million gallons of hog waste into a stream that feeds the Neuse River. State environmental officials were thankful that more spills were avoided, especially since numerous other lagoons in the east were known to be filled to capacity.

Farther inland in Raleigh, well to the west of Bertha's official track, the storm still managed to have an impact. The National Weather Service reported that around 2:00 P.M. on July 12, hours before Bertha's official landfall, a tornado touched down in southwest Raleigh and crossed into Cary. Trees were uprooted and debris was scattered, but no injuries or serious damages occurred. Another windstorm whipped across the state fairgrounds, tearing away the roof of the Hunt Horse Complex while a 4-H horse competition was under way. Almost 400 horses and 1,800 spectators were badly frightened, but no one was injured. According to Weather Service officials, both episodes were probably spawned by the approaching hurricane.

Even though Bertha, as a category two, didn't rank as a "major" hurricane, its overall impact on eastern North Carolina was nonetheless significant. Immediately after the storm, more than 400,000 state residents were without power, and some suffered without electricity for days. Utility crews poured into the stricken areas and worked long hours to remove downed trees, repair lines, and restore power. Governor Jim Hunt announced that thirty-four counties were in a state of emergency, and President Clinton later declared seventeen counties eligible for federal disaster assistance. In North Carolina, surveys indicated that over 1,100 homes were destroyed and another 4,000 damaged. The Federal Emergency Management Administration (FEMA) estimated that 750,000 residents in North and South Carolina were evacuated. American Insurance Association estimates of insured losses in the whole of the United States totaled $135 million, with most of these centered on coastal North Carolina. According to the National Climatic Data Center, a conservative ratio between total damage and insured property damage is two to one, compared to previous hurricanes. Therefore, the total estimated losses for hurricane Bertha were placed at two times $135 million, or $270 million. The Associated Press later reported, however, that Bertha caused an estimated $1.2 billion in losses in North Carolina. The U.S. Virgin Islands also suffered extensively from Bertha, although no dollar figures are available. Surveys indicated that almost 2,500 homes were damaged on St. Thomas and St. John.

As most coastal business owners know, the economic impact of a hurricane strike stretches far beyond just the costly repairs to roofs, decks, and piers. Bertha was especially unkind to those businesses that thrive on tourism, since it happened to fall in July—the peak month of the year for sales and property rentals. The storm upset the plans of thousands of beach-bound vacationers, who were forced to cancel plans or look for other destinations. Studies at the N.C. State University Department of Parks, Recreation, and Tourism Management suggest that the cost runs to about $6 million a day for every day that the state's beaches are empty. And the lingering economic effects are sometimes hard to quantify. Most coastal retailers know, however, that when the cash registers are quiet during the summer, it makes for a long, long winter.

Bertha was responsible for two deaths in North Carolina, both from automobile accidents related to the storm. An eighty-four-year-old woman was killed in a two-car wreck in Hillsborough, and another woman died in a mishap at Nags Head. Overall, Bertha directly or indirectly claimed twelve lives in the U.S. and Caribbean, including a four-year-old boy in Pensacola, Florida, who was killed when an F-16 jet crashed into his home. The jet was one of forty-nine planes being moved from Shaw Air Force Base in South Carolina to Eglin Air Force Base near Pensacola. Other casualties included the death of a teenager in an automobile accident in New York, a surfer who drowned off the New Jersey coast, and four deaths in Puerto Rico and St. Martin caused by automobile accidents, drowning, and electrocution.

After all the media attention devoted to the tumultuous hurricane season of 1995, most storm watchers along the coast were poised for whatever the 1996 season might bring. But Bertha still caught some by surprise. For a whole new generation of residents across eastern North Carolina, this storm will be remembered as their first hurricane. Others in recent years, such as Emily, Hugo, Gloria, and Diana, affected various portions of the state, but Bertha was the first hurricane since Donna in 1960 to have a broad impact across the eastern counties. As a rare July storm, it struck in the heart of the tourist season and had a lingering affect on the coastal economy. But most incredibly, after such a long period with so few hurricanes, coastal residents soon found themselves battening down for another major storm when hurricane Fran struck the same portion of the coast just eight weeks later.

FRAN (SEPTEMBER 5–6, 1996)

Throughout the storm my family was drawn closer and closer to one another because there was no TV or radios to use as excuses to get out of spending quality time together. . . . We had no electricity or water. We did have a gas stove, so we could cook stuff like ravioli. We ate a lot of that! It seemed almost like Little House on the Prairie. *—Ashley Welker, from* Fran, *a 1996 collection of essays from Topsail Middle School's seventh-grade students*

In the weeks following hurricane Bertha, residents along the North Carolina coast who had been affected by the storm did their best to repair their damaged properties and bring their lives back to normal. Trees were cut and

After hurricane Fran, some of the most remarkable scenes of destruction were found on Topsail Island. The storm's surging tide severed the island in numerous places, washing away homes, roadways, and utilities. (Photo by Randy Davey; courtesy of the Jacksonville Daily News)

hauled, signs were replaced, shingles were patched, and docks and boats were repaired. In heavily damaged communities like North Topsail Beach, the recovery was not so simple. There, state officials wrestled with controversial questions about rebuilding on the fragile island. But overall, most coastal communities were bouncing back nicely—they had to if they wanted to capitalize on what was left of the lucrative summer season.

But then, just as the kids went back to school and the Labor Day holiday approached, more talk of hurricanes spread across the airwaves. First it was Edouard, an impressive category four that was the strongest cyclone of the season. Fortunately, as this storm moved well northeast of the Caribbean Islands and toward the U.S. mainland, it was deflected by a trough over our coast, causing it to turn northward. Edouard passed midway between Cape Hatteras and Bermuda on September 1 and eventually tracked just off the coast of New England. But unfortunately, a few days behind Edouard came another storm whose course would not be diverted—hurricane Fran.

Fran was a classic Cape Verde hurricane that lived and died during the heart of the hurricane season. Its origins were familiar—out of Africa as a tropical wave with a typical westward movement—but its early development may have been thwarted somewhat by the presence of the large and powerful Edouard some 750 miles to the west-northwest. After finally reaching tropical-storm strength and earning its name on August 27, Fran literally traveled in the wake of Edouard, passing northeast of the Lesser Antilles with little impact on the Caribbean. Then deep convection became more concentrated, and Fran's winds reached hurricane strength on August 29. By the following day, however, it was downgraded to a tropical storm, possibly due again to the presence of Edouard. Fran slowed in forward speed, turned toward the northwest, and for the moment seemed too weak and distant to pose a serious threat to the Carolina coast.

As Edouard tracked north on August 31 and became less of an influence, Fran regained hurricane strength and began to accelerate. The subtropical ridge that had once kept Edouard at sea was now itself over the Atlantic and helped to steer Fran west-northwest, on a course then roughly parallel to the Bahama Islands some one hundred miles to the east. Over the next three days the hurricane strengthened consistently, finally reaching category-three intensity, with maximum sustained winds of 115 mph, on September 4. At this point, forecasters looked to the next nearby weather system that could influence the hurricane's course—a mid-to-upper-level low-pressure system drifting over Tennessee.

Forecasters and residents along much of the southeast coast were now closely monitoring the hurricane's every move. On September 4, the National Hurricane Center issued a hurricane watch from Sebastian Inlet, Florida, to Little River Inlet, South Carolina, based on their impressions of its forward direction. Late that evening, while poised some 250 miles off the Florida coast, Fran reached peak intensity. Its minimum central pressure dropped to 27.91 inches, and maximum sustained winds reached 120 mph. By this time, hurricane warnings were in effect for much of South Carolina's coast and all of North Carolina's, including the Albemarle and Pamlico Sounds. Everyone now acknowledged that Fran was a dangerous storm. Thousands turned their attention to the computer-generated landfall probabilities displayed on the nightly news and the World Wide Web. No one knew exactly where the storm would strike, but it was frightfully clear that landfall was imminent.

Full-blown evacuations got under way along the beaches of North and South Carolina, based on an expected landfall near Myrtle Beach. Throughout the morning on Thursday, September 5, a stream of cars fled inland, away from the "danger zone" near the coast. As the storm drew closer to the Carolinas, it gradually appeared that landfall would occur a little farther to the north than forecasters had earlier predicted—and it did. Fran came ashore near Bald Head Island around 8:30 P.M. Thanks to good early warnings and the storm's lack of rapid acceleration before landfall, most residents who had wanted to leave the beaches had ample time to do so.

At the time Fran made land, it was still a minimal category-three hurricane, with maximum sustained winds of 115 mph and an estimated minimum pressure of 28.14 inches. The strongest winds were believed to have run in streaks within deep convective areas of the circulation, north and northeast of the center. And Fran was no small storm. Before landfall, reconnaissance reports indicated that hurricane-force winds were measured up to 145 miles away from the eye. Before their televisions were darkened by the loss of power, most residents in the hurricane's path were able to watch the ominous swirl of radar images as Fran advanced over them. By the time it moved inland, the storm appeared to have swallowed up much of the state.

Fran's course over land seemed almost to follow the Cape Fear River, affecting a broad region that extended upstate. Because the hurricane's greatest power was concentrated on its eastern side, the beaches of New Hanover, Pender, Onslow, and Carteret Counties were hammered with the worst winds and tides. To a large degree, this was the same region that had suffered during Bertha just weeks before, but Fran was different. It was a larger, more intense storm, and its impact along most of this stretch of coast was far greater than that of the July storm.

Fran's course carried it deep inland, up the Cape Fear, west of I-40, and into the heart of the Raleigh–Durham–Chapel Hill Triangle. Along the way, its howling winds battered every county in its path, felling millions of trees and knocking out power over a major portion of the state. According to National Weather Service reports, the center of the storm tracked through portions of Brunswick, New Hanover, Pender, Sampson, Harnett, Johnston, Wake, Durham, and Person Counties before passing into south-central Virginia. Because of the storm's strength and interior course, damages were heavy all along its path. Destructive winds and flooding rains tore through the region, wreaking havoc reminiscent of Hugo's visit to western North Carolina in 1989. Gusts above hurricane strength were felt as far inland as Durham, and isolated bursts above 75 mph may even have reached into Virginia.

Officially, hurricane Fran was downgraded to a tropical storm around 3:00 A.M. on September 6, shortly before its center reached the Raleigh area. It continued to lose energy as it passed into Virginia, where it was downgraded further to become a tropical depression. It still had a significant impact on the Virginia countryside, however, as torrential rains caused flooding in the region near Danville and South Boston. As the storm slowly continued northward over the eastern Great Lakes, it gradually lost its warm central core and was no longer tracked by hurricane watchers. On the evening of September 8, Fran became extratropical while centered over southern Ontario, Canada. With little fanfare, its remnants were later absorbed into a passing frontal system.

From a meteorological perspective, Fran was one of the most intensely studied hurricanes in history. Scientists used a wide range of instruments and techniques to scrutinize the storm on land, at sea, and in the air. These studies yielded reams of data that may ultimately improve some basic understandings of hurricane behavior. Perhaps one of the most interesting experiments was conducted aboard the NOAA "hurricane hunter" aircraft. Reconnaissance flights have been flown into hurricanes for decades, but always while the storms were at sea. Fran marked the first time that researchers had ever flown

through a hurricane that was moving over land. "We've never done it before because it was considered too dangerous," said NOAA researcher Mark Powell in an interview with the *News and Observer*. "The water is fairly smooth, so the levels of turbulence aren't as intense as they are over land. No one knew what to expect."

The mission flew for three hours after landfall, looping over Raleigh and Durham around 10:30 P.M., then turning around and passing through the advancing eye south of Goldsboro. The aircraft was loaded with various weather instruments, including Doppler radar and special devices to measure the storm's barometric pressure, winds, and rainfall. Powell indicated that the mission was successful in providing "the most detailed picture we've ever had of what happens in a hurricane after landfall." In total, the reconnaissance flights during Fran (all of which were conducted by the U.S. Air Force Reserves), made 71 center fixes in the course of 17 flight missions.

Other scientific studies were under way as Fran roared ashore. Buoys loaded with weather instruments were deployed off the Carolina coast in advance of the storm. This practice has become somewhat routine in recent years, but Fran apparently was the first storm in which the buoys were in position to report from the eye of a passing hurricane. Also, scientists from the National Severe Storms Laboratory—the same government lab featured in the movie *Twister*—were in Wilmington when Fran arrived. These scientists normally focus their studies on severe thunderstorms and tornadoes; now, for the first time, they were present to observe the conditions of a landfalling hurricane.

And the scientists had plenty to observe. Wind instruments all over eastern

North Carolina were tested by the storm. Unfortunately, a few were destroyed, and some made recordings that were later discounted because the instruments either were mounted at an improper elevation (the World Meteorological Organization standard is ten meters, or thirty-three feet, above the ground), or simply were not reliably accurate. Nevertheless, the National Weather Service was able to confirm wind reports that included *gusts* of 124 mph at Frying Pan Shoals Light, 122 mph at Figure Eight Island, 105 mph in Southport, 105 mph in New Bern, 100 mph in Greenville, 92 mph at the Duke Marine Laboratory in Beaufort, 80 mph at Seymour-Johnson Air Force Base in Goldsboro, 79 mph at Raleigh-Durham International Airport, and 86 mph at Efland, west of Durham. There were, after the storm, published but unconfirmed reports of gusts to 137 mph from Hewletts Creek in Wilmington and 126 mph at Wrightsville Beach. As is usually the case with landfalling hurricanes, reports of *sustained* hurricane-force winds were difficult to find, although a reading of 90 mph was recorded at Frying Pan Shoals (based on a two-minute averaging period). The lowest barometric pressure observed on land was 28.14 inches, recorded in Southport at the time of landfall and communicated to the National Hurricane Center by amateur radio volunteers.

Wilmington's Doppler weather surveillance radar measured winds aloft in excess of 138 mph as the inner convective bands of the storm approached the Cape Fear area before landfall. In addition, reconnaissance flights crisscrossing through the storm recorded wind extremes, although these recordings were made at high altitudes. Flight-level maximum winds of 130 mph were measured fifty miles east of the storm center just before landfall. On the previous day, radar screens on board the plane had indicated that the eye was about twenty-five miles in diameter, but surveillance indicated that the core of the storm weakened and the eye wall opened during the last few hours before it moved onshore.

According to National Weather Service reports, the heavy storm surge that affected much of North Carolina's southeast coast ranged from eight to twelve feet. Still water marks measured inside buildings provided these elevations; exterior water levels were generally higher due to wave action. The highest surges were observed at Carolina Beach, Wrightsville Beach, Figure Eight Island, and Topsail Island. Some longtime residents at Wrightsville Beach compared the flood level during Fran with that of hurricane Hazel in 1954 and found that Fran's was several inches higher. Hazel was a much more intense storm, but it struck some forty miles farther down the coast. Preliminary reports from the Weather Service indicated that Fran's surge was ten feet at Swansboro and New Bern, nine feet at Washington and Belhaven, and seven feet at Atlantic Beach. Still water surge heights of five feet were measured in Southport and Beaufort. A large portion of the dunes between Figure Eight Island and Emerald Isle had been washed away by Bertha, setting the stage for

Millions of trees fell across the state during the passage of Fran, and thousands crushed homes and cars as they fell. At Camp Lejeune, parked cars were hammered by fallen pines. (Photo by Randy Davey; courtesy of the Jacksonville Daily News)

extensive beach erosion and ocean overwash during Fran. With little dune structure left to protect these areas, the surging tide and breaking waves demolished homes, undermined roads, and covered streets and yards with deep sand and debris.

The destructive floods that hit many portions of the state did not just sweep in with the storm surge, but also fell from the skies. Drenching rains bore inland as Fran came ashore, with the heaviest bands of precipitation very closely following the storm track. To make matters worse, some portions of the state received monsoonlike rains even before Fran approached the coast. On September 3, two days before the hurricane's landfall, Carteret and Onslow Counties were inundated with rains measured at up to 7.5 inches in some locations. These rains filled creeks and ditches and saturated soils near the coast. At the opposite end of the state, strong thunderstorms associated with an upper-level low poured incredible rains on several western communities just before the hurricane struck the coast. The area around Bat Cave, Lake

Lure, and Chimney Rock received up to 11 inches of rain in just three hours; Sugar Loaf Mountain recorded 12.49 inches. As a result, about seventy homes and businesses were severely damaged or destroyed. And in the central part of the state, the Triangle and northern Piedmont had about 4 inches of rain in the three days immediately preceding Fran. These rains also saturated the ground and set the stage for the record-breaking floods that followed.

Rainfall totals exceeding 6 inches were common along the path of the hurricane as it moved over land. Wilmington radar precipitation estimates ranged as high as 12 inches over portions of Brunswick and Pender Counties, and the Newport Doppler system estimated similar amounts over Duplin and Onslow Counties. Cherry Grove Pier in South Carolina reported 8.36 inches, Pope Air Force Base in Fayetteville 6.72 inches, and New River Air Station at Jacksonville 7.05 inches.

As Fran tracked inland, its center of circulation held together longer than expected, partially due to its large size. Fran was as large as Hugo and larger than Andrew, even though it wasn't as powerful as either of those storms. Its central core finally collapsed just southeast of Raleigh, almost directly over the city of Garner. These factors spelled disaster for the Raleigh area. The *News and Observer* described the result: "Two large bands of intense winds and rain surrounding the center suddenly rushed together as the storm collapsed inward on itself over Raleigh. That caused the hurricane to drop its heaviest rains on the city." Though reports suggested that up to 9.5 inches fell at some Triangle locations, the Raleigh-Durham airport officially recorded 8.80 inches, still enough to shatter the old twenty-four-hour record of 6.66 inches set back in September 1929. (Hazel brought only 4.93 inches to the Triangle.) Other records marked at the airport included the highest recorded wind gust (79 mph, which beat the previous record of 64 mph set during a thunderstorm in May 1996) and the lowest barometric pressure for September (28.84 inches). A slightly lower pressure reading of 28.75 inches had been recorded at the airport during Hazel in October 1954.

The worst of the rains were yet to come, however. Fran continued to dissipate as it moved into Virginia and dumped its heavy clouds over much of the state. Some of the most impressive precipitation occurred near the Shenandoah National Park, where the rainfall amounts were enhanced by the orographic effects of the central Appalachian Mountains. The highest recorded amount was 15.61 inches, though numerous stations measured in excess of 14 inches. Heavy rains also spilled into portions of West Virginia and Pennsylvania, where flooding was considered the most severe in years.

In addition to monitoring estimated rainfall over the areas affected by Fran, Doppler radar stations also indicated the presence of several tornadoes in North Carolina and Virginia. According to the National Weather Service, confirmation of such reports was difficult due to the extensive nature of straight-

After riding out hurricane Fran at Carolina Beach, Paula Cauble checked on the first-floor damage of her neighbor's home. (Photo by Robert Willett; courtesy of the Raleigh News and Observer)

line wind damage across the region. Some residents near Hampstead reported hearing the distinctive "freight train" sounds of a twister barreling through their neighborhood, though so many trees were downed throughout the area that isolating a particular path of destruction was tricky. Weather experts agree, however, that funnel clouds were very likely present in some of the storm's stronger convective bands.

Meteorologically, Fran was a major hurricane—a real category three with all the trimmings. But perhaps even more impressive than the record-breaking wind, storm surge, and rain were the scope and severity of destruction the storm laid across the land. Decades had passed since the Tar Heel state had felt this kind of blow, and its force painfully exposed our vulnerabilities. Not only did the storm shatter property wherever it went, but it also altered the lives and challenged the spirits of its victims.

As Fran came ashore near the mouth of the Cape Fear River, residents in nearby Southport were expecting the worst. Though well informed about the storm's anticipated intensity, many older inhabitants still remembered the last benchmark hurricane to hit the state—hurricane Hazel—even though it had landed more than four decades earlier. They knew, perhaps better than most, that Fran should be taken very seriously.

Fortunately, Southport and the Brunswick County beaches were spared the worst of Fran. Even though the storm's eye passed directly over Southport, wind damages there were generally lighter than expected. Slight damages were reported to roofs, signs, and docks, and some of the town's trademark live oaks were toppled. The nearby Brunswick Nuclear Power plant, which supplies 20 percent of the Carolina Power and Light Company's (CP&L) electricity, was shut down sixteen hours before the storm arrived, and no damages were reported there. Ironically, while the inland areas of New Hanover and Pender Counties were buffeted by high winds, an eerie quiet fell over the streets of Southport. The calm lasted so long that some residents ventured into the streets to peer up at the twinkling stars, despite warnings to stay put. After the hurricane had passed, Brunswick County emergency management director Cecil Logan told *USA Today*: "We stayed in the eye for more than two hours, which is unheard of. We stayed in that eye, and the storm tore up everyone else."

At Long Beach, more than two hundred homes received some damage, and nineteen were rendered uninhabitable. Many were oceanfront cottages whose septic systems were undermined or damaged by the storm tide. The Long Beach Pier lost about a hundred feet of its length, but other piers on the island were relatively unharmed. At Holden Beach, seven homes reportedly "fell into the ocean," and several feet of dunes were washed away. At Ocean Isle, the septic systems of fourteen homes were damaged, and several cottages were flooded by overwash. Overall, though, these same beaches that were swept by Hazel's awesome seventeen-foot surge fared very well with Fran. Because they were on the weaker side of the eye, the storm surge was moderate, measuring only about six feet at Holden Beach. In the weeks following the storm, tourists who were forced to cancel vacations farther north jammed the county's beachfront hotels.

Around Cape Fear, it was a different story. The beaches of New Hanover County caught the brunt of the hurricane and were left with a massive mess. Fort Fisher, Kure Beach, Carolina Beach, Wrightsville Beach, and Figure Eight Island all were exposed to Fran's destructive core. As high winds peeled back roofs and snapped utility poles, the surging Atlantic sucked away dunes and poured into the streets. The tide filled some homes with water and sand and radically displaced others. In Carolina Beach, one house floated off its foundation and came to rest on Canal Drive, some two hundred feet inland. The *Wilmington Star-News* reported that the town was under six feet of water at times. Other stories held that cars were floating about during the peak of the storm, bumping into telephone poles and buildings. As the flood worsened, rising water forced the town's police department to move from its headquarters on Canal Drive to the town recreation center. One resident, who believed pounding waves were about to demolish his condominium building, called 911 at the height of the storm to relay information to his next of kin. Some longtime residents emphatically insisted that the damage was "just as bad as Hazel." In all, Carolina Beach had over 930 damaged homes, 41 more destroyed, 75 damaged businesses, and over $33 million in private property losses.

At nearby Kure Beach, the scene was similarly depressing. Dozens of ocean-front homes and condos, already compromised by Bertha a few weeks earlier, suffered major structural damage. Several feet of sand were carried inland by the surging waters and deposited in first-floor rooms of vacation resorts. Porches, decks, and awnings disappeared, either swept away by wind and water or buried out of sight. Debris from damaged structures drifted in the flood, covering streets and lawns with a potpourri of lumber, appliances, household furnishings, and broken glass. A portion of Atlantic Avenue washed away, and other streets were barely recognizable. Water and sewer lines were ruptured, 68 businesses were damaged, and more than 260 houses were destroyed. Kure Beach was near ground zero during Fran, and few communities faced more significant destruction.

One well-known facility affected by the storm was the North Carolina Aquarium at Fort Fisher, which suffered only minor structural damage but was forced to rely on its diesel-powered generator to keep air and water flowing to its display tanks. After two days, fuel began to run low, and curator Paul Barrington and his staff faced the possibility of having to release or lose their collection of fishes that had taken years to acquire. But fortunately, after reading about the aquarium's dilemma in the *Wilmington Star-News* on September 8, the wife of a CP&L worker urged her husband to make the aquarium a priority. By 11:00 A.M., the electricity was back on again, and the fishes were saved.

Up the coast at Wrightsville Beach, a storm surge of close to eleven feet tore down dunes and flooded expensive beach houses. Almost the entire town was underwater at one point. When residents at Harbor Island finally returned to their homes, some found scum lines on their walls four feet above the floor. Up and down the beach, deep sand covered Lumina Avenue, mixed with debris from first-floor cottages and apartments. Beach chairs and washing machines were not the only losses — some people lost personal keepsakes such as wedding albums, home videos, love letters, and music collections. Virtually anything left at ground level was either saturated by rising salt water or washed into the streets. At the Bridge Tender restaurant near the bridge to Wrightsville, witnesses reported that high water "floated bottles off the bar." On the northern end of the beach, a six-foot dune was flattened, edging Mason Inlet even closer to the imperiled $22 million Shell Island Resort. Like the other nearby beaches, Wrightsville suffered heavy damages: more than 560 homes damaged, 13 houses destroyed, and 50 businesses damaged.

In Wilmington, it was largely Fran's nasty winds that caused trouble. Bertha had already thinned out the weaker trees, but countless more either snapped above the ground or heaved over, roots and all, during Fran. Power lines and traffic signals were torn and tangled, gas station canopies buckled and fell, and residential roofs were stripped. Empty tractor trailers were blown on their sides, and commercial buildings had their metal roofs peeled away. One of the casualties was the Days Inn, which lost part of its roof during the storm, to the astonishment of guests who had retreated to their rooms on the top floor. Around the city, evidence of Fran's visit was easy to spot on the morning after the storm.

Few landmarks in Wilmington got more attention after the hurricane than the historic First Baptist Church on Market Street, which lost its steeple to one of Fran's early gusts. Sometime after 10:00 P.M. on the night of the storm, Rev. Mike Queen was summoned to his church to inspect the damage. He arrived during the quiet of the hurricane's eye and found that a major portion of the church's 197-foot spire had exploded onto the sidewalk below. The

As Fran's tides reached their peak near Wrightsville Beach, large boats from the Intracoastal Waterway were lifted onto Airlie Road, carrying with them broken sections of dock. Fran caused an estimated $50 million in losses to boats in North Carolina. (Photo by Robert Willett; courtesy of the Raleigh News and Observer)

copper-topped steeple, built between 1860 and 1870, was thought to have been used as a lookout post during the Civil War. It had since withstood many hurricanes, including Hazel in 1954 and the 135-mph winds of hurricane Helene in 1958. After studying the mountain of bricks beside his church, Queen proclaimed that the spire would be rebuilt. He also described to the Associated Press his perspective on the loss: "God created a world that allows nature to do what it will."

Curator Harry Warren of the Cape Fear Museum summarized the First Baptist Church's calamity thus: "It came down like thunder. Then, it ended up being symbolic of the entire hurricane experience in Wilmington. The next day, every news cameraman in the free world was out there filming in front of it. Satellite trucks were all over the place and these news reporter guys were dressed up like something out of Banana Republic. It was really quite a scene."

Emergency management officials later reported that one-quarter of Wilmington's homes were damaged. According to the *News and Observer*, 9 businesses sustained major damage, 320 homes were destroyed, 530 homes suffered major damage, and more than 3,000 had minor damages. As in most other communities in the area, power was out across the city. Mayor Don Betts estimated the impact on public buildings to be at least $5 million and described Fran as a "major, major storm."

Across New Hanover County, boat owners had done their best to prepare for the storm, some harboring their boats in protected creeks, others trailering them inland. Many, though, underestimated the power unleashed by Fran. Along docks from Snow's Cut to Bradley Creek, boats small and large were tossed, beaten, and sunk. Like toys in a bathtub, many broke free of their moorings and rode the tide to high ground. After the waters receded, Masonboro Road was littered with broken docks and misplaced boats, just as it had been forty-two years earlier after Hazel. One vessel tossed onto high ground was the *Hurricane Rock*, which was selected as the backdrop for Tom Brokaw's poststorm broadcast of the *NBC Nightly News*.

At Bradley Creek, several large boats and the floating dock to which they were tied broke loose during the storm and smashed into the Oleander Drive bridge. Boats at the nearby marina had suffered a similar fate after Bertha. The Scotts Hill Marina was also beaten by Fran's pounding storm surge, sending sections of dock and expensive pleasure crafts into the nearby woods. Before Fran hit, a local film studio had scheduled to shoot scenes at the marina for the movie *Buried Alive II*. Afterward, producers agreed to foot half of the bill for the dock repairs so that filming could proceed. According to the *Wilmington Star-News*, losses from damaged and destroyed boats alone totaled $50 million.

Perhaps the most amazing destruction inflicted by Fran fell upon the hapless communities of Topsail Island along the Pender-Onslow coast. Topsail Beach, Surf City, and North Topsail Beach were all slammed hard by hurri-

The 197-foot spire of Wilmington's historic First Baptist Church fell victim to Fran's gusts during the early part of the storm. The massive brick steeple, which had withstood numerous hurricanes over the years, was blown over into the street. (Photo by Robert Willett; courtesy of the Raleigh News and Observer)

cane Bertha in July, only to catch the worst of Fran mere weeks later. Bertha caused extensive property damages, but more important, it washed away the island's protective dunes. Afterward, very little remained to slow the ten-foot storm surge that accompanied hurricane Fran. Across most of the region, everyone acknowledged that the second storm was more serious. After Fran had passed, Onslow County public works director Dave Clark compared the two hurricanes for the *Wilmington Star-News*: "We had a little breeze come by here on July 12th. This was a *real* storm."

The devastation at Topsail was immense. Nearly all of the front-row cottages in Topsail Beach were destroyed, and about half of the second-row homes were either destroyed or heavily damaged. Flights over the island on the morning after the storm were the only means of assessing the scope of the loss. Along one three-mile stretch of N.C. 50 in Surf City, the *News and Observer* estimated, about 200 out of 500 homes had "obvious roofs and walls missing, foundations crumbled, or windows blown in." In many locations, splintered stubs of pilings and half-exposed septic tanks were the only recognizable features that remained where houses had once stood. More than 300 homes incurred damages that exceeded half their value. Added to the debris on the beach were hundreds of car tires torn loose from an artificial reef about five miles offshore.

North Topsail was also hit hard again. It was estimated that more than 90 percent of the structures in town were either destroyed or damaged beyond repair. Many had never been fully restored after Bertha, but others had been completely remodeled. Some unfortunate homeowners had literally just finished repairs during the week of Fran's arrival. After Fran, the town hall and police station were gone, and roadways and utilities that had been patched together after Bertha were demolished. The tide that swept over the island was so powerful that it lifted entire cottages and floated them hundreds of yards into the marsh. Its force was so strong, in fact, that it carved six new inlets across Topsail Island, slicing up the beach road and isolating entire communities.

Amazingly, scores of residents remained in their island homes as Fran came ashore, though not all of them wanted to be there. Several said they simply waited too late to leave and found themselves trapped once waves pushed over the beach road. A few stayed because they wanted to "protect their property," and others said they stayed for the thrill of it. Some who rode out the storm in their elevated homes described the ordeal of listening to the thundering surf break under their feet. They felt their cottages shake and heard the waves wash through their ground-floor garages.

One of those who hadn't planned to stay but waited too long to leave was Len Gioglio. Earlier that day, Gioglio sent his family to a nearby shelter while he remained in his North Topsail Beach cottage to make storm preparations. He knew he had waited too long when he saw deep water covering New Inlet Road, the only escape route off the northern end of the island. His story was told in the *Wilmington Star-News*:

> As the waters rose Thursday night around his home, Len Gioglio did the only thing he could think of: He tied a rope around himself, lashed it to his battered home, and held his dog in his arms for two terrifying hours as they were slammed by Hurricane Fran.
>
> Each time the 4-foot storm surge slammed them down the rope, Mr. Gioglio, 37, pulled himself and his dog back up the rope, thinking that the next time they might not make it.
>
> "I thought I was going to die," Mr. Gioglio said Sunday, his voice shaky. "You just couldn't imagine how strong the wind was. . . . It rained so hard you couldn't see a few inches in front of you."

Beyond the obvious destruction of manmade structures on the island, the natural environment was also tattered after the storm. Heavy beach erosion had moved mountains of sand, wiping out what remained of the dunes and their fragile vegetation. So much construction debris, paper, plastic, glass, appliances, and toxic household products was washed into the marsh or buried in the sand that the area might never be completely cleaned up. Local conservationist Jean Beasley was quoted in the *North Carolina Herpetological Society Newsletter* as saying that Topsail Island was "an ecological disaster, with tons of treated lumber and untreated sewage in the ocean and sound, mosquito trucks spraying every night, and those snakes that had survived the storm being killed by the hundreds as they sought high ground. Every sea turtle nest on the island had been inundated."

Farther up the coast, the town of Swansboro was still picking up the pieces from hurricane Bertha when Fran came ashore. Docks that had just been rebuilt were washed away again, and more stately trees were toppled into the streets. Winds and tides battered dozens of homes and shops along the waterfront. Casper's Marina was hit especially hard, and the Snap Dragon restaurant was severely damaged. The Christmas House, a waterfront shop filled with holiday ornaments, was floated twenty feet off its foundation and flooded with five feet of water. By the time owner Henry Schindelar arrived at the scene the following morning, looters had already made off with gifts and decorations. Schindelar told the *News and Observer*: "It's a mess. I don't know whether to clean it up or just bring in the bulldozer."

In Carteret County, tides along the beaches of Emerald Isle and Pine Knoll Shores caused extensive beach erosion, and soundside docks were washed away. At the east end of Emerald Isle, sand and water covered about ten blocks of Ocean Drive, and several homes were torn open by the winds. Septic tanks were exposed, and broken glass and pink insulation spread across the dunes. Initially, Emerald Isle officials thought as many as thirty-six homes might be lost, but eventually all but six were repaired. Winds ripped awnings, porches, and decks and stripped away shingles on countless cottages. East of Beaufort, however, the damage lessened, and most residents felt lucky after viewing the aerial videos of the beaches farther south. But Fran did have a lingering effect on the forests of Atlantic Beach, Pine Knoll Shores, and Emerald Isle. Like other portions of the coast affected by the storm, the woods on Bogue Banks suffered a severe infestation of southern pine beetles during the summer of 1997 that wiped out thousands of mature loblolly pines. The voracious insects capitalized on the effects of the previous year's storms, first attacking pines weakened by salt spray and wind damage. The infestation ultimately destroyed most of the island's large pines, in many cases costing landowners more in tree removal expenses than they had paid for other, more traditional hurricane repairs.

Riverfront communities not in the storm's direct path were nonetheless soaked by flooding tides and heavy rains. Downtown New Bern was submerged by the rising waters of the Neuse, which also undercut the roadway on the U.S. 17 bridge, causing a six-foot section of pavement to collapse. The damaged road forced officials to close the bridge for ten hours. In Belhaven, town manager Tim Johnson reported that 30 to 40 percent of the houses in town were damaged by floodwaters from the Pungo River, which rose 8.5 feet above normal. In Washington, about three-quarters of the downtown area was under water at the height of the storm. Residents described the streets as "waist-deep rivers." The combination of floodwaters from the Pamlico River and heavy rains draining from nearby creeks left several blocks of the city under water more than a day after the hurricane's passing. Even in Elizabeth City, Fran's winds pushed water down the Pasquotank River, flooding the town's harbor, wrecking docks and piers, and sinking a forty-two-foot sailboat.

In Jacksonville and surrounding Onslow County, residents endured the eastern edge of Fran's dangerous core, and damages from wind and flooding were widespread. Heavy rains earlier in the week had already filled ditches and creeks, and by the time Fran pushed inland, the New River had spread more than a quarter of a mile beyond its banks. Tall pines snapped at house-top level, crushing scores of cars and homes under fallen timbers. Propane tanks floated into the streets, tractor trailers were bowled over, and telephone poles leaned crazily. Last-minute evacuations were required in some locations. Jacksonville police detective Candido Suarez reportedly commandeered a Jet-ski and rescued more than twenty people from a flooded street. A city water tanker and surplus military truck carried others to high ground, including dialysis and heart patients.

Much like Hugo in 1989, hurricane Fran first wrecked communities along the coast, then marched inland with remarkable force. Its center tracked just west of I-40, blasting the mostly rural counties between Wilmington and Raleigh with hurricane-force gusts and hefty rains. Although they never received the media attention accorded to the more populous cities or the hard-hit resort beaches, many small rural towns in Fran's path were badly pounded by the storm. Downed trees, washed-out bridges, and flooded roads isolated numerous communities, slowing rescue efforts and making recovery difficult. Among the eastern highways blocked or closed were portions of N.C. 53, 50, 11, 41, 210, 903, 904, 905, and 411, U.S. 421, and dozens of other state roads. Concerned over the rise of floodwaters, the state highway patrol also closed portions of Interstates 40 and 85 on Friday after the storm.

In Bladen County, downed trees and pouring rains tested the skills of rescue teams called out during the hurricane. An expectant mother put in an emergency call just after midnight, when ambulances had ceased operations due to the storm. But because her baby was breech, emergency workers decided to attempt the transport. Calls to hospitals and military bases for a rescue helicopter were useless because high winds kept the choppers grounded. Risking their own lives, four different crews of emergency personnel headed toward the anxious mother from different directions, each battling the gusting winds and treacherous conditions. Firefighters used chain saws to cut trees along the way, clearing a path for the ambulances to reach the mother's home.

So many trees were downed that it took fifty-three minutes for rescuers from Kelly to travel just ten miles. Eventually, they reached the anxious mom and took her to the county hospital, where the baby was delivered. Mother and child were both reported in good condition.

In rural towns like Warsaw, Wallace, Burgaw, and Clinton, streets were littered with massive tree branches, loose shingles, and broken glass. Commercial signs were buckled and shattered, and misplaced power lines were draped over storefronts like Sillystring. Warped pieces of tin hung from trees across the countryside, having been stripped from the roofs of farmhouses and barns. Some yards were filled with the trunks of massive oaks and huge earthen cavities where the trees had once stood. The *News and Observer* reported that "virtually every road in Sampson and Duplin Counties was blocked by fallen trees, utility lines, swollen rivers, or all three." In Kenansville, a strong gust lifted away the copper dome of the Duplin County courthouse, built around 1911, and dumped it into the street below. The busted shell was later retrieved by a sheriff's deputy.

Fran cut right through the heart of North Carolina's eastern farm belt, and the resulting damages were enormous. As the slowly collapsing eye danced along the Sampson-Duplin county line, gusting winds peeled back the roof of the UAP/Carolina agricultural warehouse, collapsing the structure's cinderblock walls onto soggy stacks of fertilizers, herbicides, and other agricultural chemicals. State officials soon showed up to check for any environmental hazards the damaged contents might pose.

In July, hurricane Bertha had dealt a severe blow to the state's farmers, causing an estimated $189 million in agricultural losses. Fran's impact was much greater and was spread over a broader region. Thousands of acres of crops were blown down or flooded, and heavy losses were reported to livestock and poultry. Especially hard hit were farmers who grew corn, cotton, soybeans, and tobacco. Tobacco farmers lost plants in the fields, to be sure, but many also lost leaves that had been harvested but not cured. The hurricane knocked out the power to drying equipment in their barns, causing the leaves to sour within days. High winds and flooding were also responsible for the deaths of more than one million turkeys and chickens, sixteen thousand hogs, and more than four hundred head of cattle. In some cases, entire flocks of poultry were lost when their buildings were flooded. The North Carolina Department of Agriculture estimated the total loss to farming at $684 million, including more than $185 million in tobacco, $55 million in cotton, and $54 million in corn.

In the hours preceding Fran's landfall in North Carolina, thousands of vacationers and residents along the coast packed their cars and fled inland. Some remained near the coastal area, staying with family or friends, and some sought refuge in public shelters that had been set up in schools and government buildings. Others drove farther inland and rented any available motel room they could find. Even in communities far from the coast, hundreds of families went to local shelters for protection from threatening winds and rising water. The American Red Cross operated 134 shelters in 51 counties, housing 9,426 people on the night of the storm. In all, according to National Weather Service reports, almost a half-million residents in North and South Carolina evacuated during Fran. But unfortunately, many who left the beaches fled inland to wait out the storm in one place they surely thought would be safe — the Triangle region of Raleigh, Durham, and Chapel Hill.

But Fran hit the Triangle hard. Not since Hazel's visit in 1954 had the Raleigh area been subjected to the kind of flooding and wind damage it endured through Fran. Most residents in the region began the evening by watching multicolored radar and satellite images spin across their television screens, knowing that the storm's path was undeniable — Fran was headed their way. As the hurricane's collapsing eye approached the region, massive trees tumbled and power lines popped and sizzled. Inside their now-dark homes, almost a million Triangle residents could only sit and listen to the whirling winds outside, waiting fearfully for the next thunderous crash of pine or oak. A few managed to sleep through the ordeal, but most paced through the night with flashlight in hand, hoping for it all to end soon.

Winds in the Raleigh area were nearly hurricane strength, and countless trees gave way to the blow. Pines typically snapped like toothpicks, whereas many large oaks fell over completely, roots and all. Hundreds of homes and businesses were bludgeoned by the falling trees, which sometimes left deep craters rimmed with muddy red clay. Downed trees and branches were everywhere — across highways, through roofs, tangled in power lines, smashed through boats and cars, in swimming pools — and virtually every neighborhood was affected. On the campus of North Carolina State University, more than 250 stately trees were lost. Those who had seen Charlotte after hurricane Hugo had instant flashbacks — the images of entire neighborhoods covered with toppled trees were eerily familiar. Raleigh — once dubbed the City of Oaks — now seemed to be the city of *fallen* oaks.

As the decaying hurricane swept over the city, residents huddled in darkened hallways to wait for the worst to pass. Those who felt the thundering crash of trees into their homes struggled to make sense of the invasion. Some managed to survive very close calls. Just ten minutes after North Raleigh resident Thao Do left her bedroom for the comfort of her mother's room, a large tree crashed through the roof and landed on her bed. The Dos told the *News and Observer* that they didn't even realize what had happened at first. They heard a loud crashing sound, "like a freight train going through the house." Bruce Do later said, "We didn't realize the damage until we opened up my sister's bedroom and saw the sky."

In Cary, Laura Hamo woke her two young sons, Evan and Eric, and moved them out of their upstairs rooms when the storm grew especially fierce around 1:00 A.M. Just ten minutes later, a sixty-foot oak toppled onto the Hamo house, landing a large branch on Evan's bed. Raleigh resident Patti Clinton escaped a similar disaster when her faithful beagle saved her life by whimpering and waking her just moments before a massive tree smashed the bed where she had been sleeping. Others in Raleigh were not as fortunate. Among the many deaths Fran caused in North Carolina was that of Mary Bland Reaves, who was killed when a single falling tree sliced through her mobile home on Nevada Drive.

Unfortunately, Fran's effects on the Triangle were not limited to punctured roofs, crushed cars, and streets blocked by leafy barricades. A record rainfall on the night of Fran's visit saturated soils already presoaked by earlier rains, and the result was some of the worst flooding in the region's history. Entire blocks of homes were swamped, as were countless businesses and major thoroughfares. In Durham, rising waters flooded city streets and forced hotel residents to flee in the middle of the night. In Chapel Hill, portions of Estes Drive and Eastgate Shopping Center were submerged. The Eno River spread beyond its banks and into downtown Hillsborough, flooding a county building and covering more of the city's streets than anyone could remember ever happening before.

But Wake County endured some of the region's most destructive flooding. Raleigh's Crabtree Creek spread far beyond its banks, inundating the area surrounding Crabtree Valley Mall and much of the heart of the city. The normally placid creek, a tributary of the Neuse River, rose more than sixteen feet in just fourteen hours, finally cresting at seven feet above flood stage. The result was a muddy sea that, by 4:25 A.M., filled the lower level of Crabtree Valley Mall with a foot of water and turned the nearby Sheraton Hotel into an island. Merchants in the mall later reported finding "water moccasins" among their merchandise. As the morning progressed, rescue workers broke through doors at the Sears Pet Center and rescued animals endangered by the rising flood.

Surrounded by six-feet-deep floodwaters that completely buried rows of parked cars, guests at the Sheraton found themselves stranded on the morning of the storm. Included on the guest list were numerous evacuees who had fled the coast in search of safe refuge. One of those was actor James Woods, up from Wilmington where he had been filming a forthcoming movie. Woods reportedly donned bedroom slippers and a bathrobe on the morning after the storm and helped make sandwiches for fellow victims in the hotel lobby. Also among the guests were a heart patient and a diabetic, both in need of medical assistance. They were aided by Ray Williams and Lupton Pittman, two Raleigh men who happened on the scene. After learning of the situation at the Sheraton, Williams and Pittman left, only to return soon after with wetsuits, emergency items, and a canoe.

Other parts of the city were also hit hard by flooding, which wrecked houses, stranded motorists, and set the stage for dramatic rescues. On the Saturday following the storm, the *News and Observer* ran a remarkable account of the floods in Raleigh:

Some of the worst flooding devoured much of the Forest Acres neighborhood near where Crabtree Creek runs under Wake Forest Road.

Jason Ferriss, a house painter who lives in the Forest Acres neighborhood, said he was surprised to find water up to his ankles when he was awakened by his two cats about 4:30 A.M., and got out of bed to find out what was wrong.

"I decided to get my cats to safety on the roof, but by the time I got a ladder up there, the water was up to my waist," Ferriss said.

Ferriss remained on the roof until daylight, and then found an inflatable air mattress in his home, which he began using to swim a few belongings out. He said the water in his house had risen to within a foot of the ceiling by then.

"I lost everything except what I have here," Ferriss said as he loaded three dripping photo albums, a basket of clothes and a rifle into his sister's car.

Several people who live in rentals at the end of Anne Street scrambled to their roofs as water poured into their homes. Two young men from a nearby street who were out checking for damage heard their pleas for help, so they ran home and returned on a power ski.

"They were screaming like the night," said Roger Reece, 25, who along with his friend Tom Quackenbush, 22, got three people to safety on Quackenbush's power ski. The water was about 8 feet deep, the men said. They said they had to walk on top of submerged cars to get to the house.

One man declined to be rescued from his roof, they said, because he wanted to stay there and videotape the scene once the sun rose. "That made me mad as hell," Quackenbush said. "We go to all the trouble to get them life jackets, show up on a Jet Ski, and then he wouldn't come off the roof."

Flooding along Middle Creek in southern Wake County was also severe. A Benson man who was attempting to cross where the creek overran Old Stage Road was swept downstream by the torrent. He eventually climbed into some small bushes, but the currents in the area were so swift that no boat could reach him. Sheriff's deputies later called in a helicopter from Fort Bragg to make the rescue. At the Triple W Airport in southern Wake County, two young Fuquay-Varina boys were rescued after they were swept 150 feet down a raging creek. A small boat was brought in to reach the boys, who were safely pulled from an isolated bank.

Silver Lake Waterpark, a popular recreational site in Raleigh, burst its dam during the night and emptied its waters across Tryon Road, stranding colorful bumper boats on the lake's muddy bottom. Bass Lake and Lake Raleigh also emptied out when their dams were breached. Yates Millpond near Raleigh was one of the more notable losses, mourned by generations of naturalists who had studied aquatic ecology along its banks. Other lakes, millponds, and farm ponds around the region suffered a similar fate—their waters broke through dams and were pulled by gravity toward swollen creeks that feed major rivers like the Neuse. According to a North Carolina Division of Emergency Management report, there were three major dam failures and twelve minor failures involving private facilities.

On Falls Lake, the Rolling View Marina was crippled by the storm, but not from flooding. Fran's high winds whipped boats from their moorings, sinking about twenty and scattering others across the lake. Falls Lake suffered in other ways, too. A North Durham sewage treatment plant lost power and flooded during the storm, releasing partly treated sewage into the reservoir that provides Raleigh with drinking water. Numerous other sewage treatment plants and water plants across the state were also disabled, forcing thousands of residents to heed health warnings in the days following the storm: Boil your drinking water!

Across scores of eastern counties, river flooding reached all-time highs during the week following Fran. The U.S. Geological Survey reported that by Friday, September 13, water levels on six North Carolina rivers and streams had set new records. The Flat River in Durham County, a tributary of the Eno, reached 17.3 feet; the Eno River at Hillsborough, 21.1 feet; Middle Creek near Clayton, which flows to the Neuse River, 14.3 feet; the Tar River in Granville County, 24.1 feet, and in Franklin County, 25.2 feet; Tick Creek in Chatham County, which flows into the Rocky River, 13.4 feet; and the Haw River in Alamance County, 32.8 feet. Three of these—the Tar River, the Flat River, and Middle Creek—reached the five-hundred-year flood level, according to the USGS.

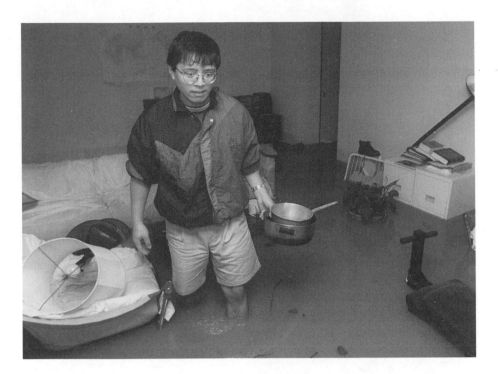

Communities downstream from Raleigh along the troubled Neuse River braced themselves for a mighty flood that took days to reach its peak. During the hurricane, floodwaters had already swamped portions of Smithfield, damaging homes and businesses along Market Street. But the worst was yet to come. Falls Lake, north of Raleigh, had risen 11 feet during Fran, reaching a new record high—262.7 feet—just 18 inches below the rim of the lake's spillway. The U.S. Army Corps of Engineers, which controls the lake's water level, feared that more rains could cause the lake to overflow or, at worst, burst its massive dam. They began a controlled release of water on September 9, dumping as much as 6,400 cubic feet per second into the Neuse by the following day. The result was an extended flood downstream, which took days to manifest in Smithfield, Clayton, portions of Wayne County, and most notably Kinston. The normal rate of release at the Falls Lake dam is 150 cubic feet per second.

After invading the low-lying neighborhoods of Johnston and Wayne Counties during the previous week, the Neuse River finally peaked in Kinston on September 17—more than ten days after Fran had swept through the state. According to the USGS, the Neuse crested in Kinston at 23.2 feet, more than 9.1 feet above flood stage, largely because of the Falls Lake discharge. The worst flooding occurred in the southeastern portion of the city and in communities along the southern side of the river. More than two hundred homes within the city limits were engulfed, and nearby mobile home parks and small hous-

Massive trees fell throughout the Triangle area, including this oak that shattered a bedroom wall on Kenmore Drive in Raleigh. (Photo by Scott Sharpe; courtesy of the Raleigh News and Observer)

es were also affected. Water was two to three feet deep in the streets of Rivermont, just south of the city. Hundreds were forced to flee as the waters crept into their yards, and many more were rescued by National Guard troops and sheriff's deputies who ran small boats through the streets. Kinston's ailing wastewater treatment plant was overwhelmed for days, releasing an estimat-

An employee of Al Smith Mazda was offered a boat ride by rescuers when floodwaters swept through the Wake Forest Road area near Hodges Street in Raleigh. (Photo by Gary Allen; courtesy of the Raleigh News and Observer)

ed four million gallons of raw sewage into the river every twenty-four hours. Already burdened with similar sewage problems upriver, the Neuse spread an unsavory odor around Kinston that was reportedly noticeable from low-flying airplanes.

Officials in Kinston, like those in other flooded towns, warned residents to avoid contact with polluted waters and to watch out for venomous snakes seeking high ground. Numerous city streets and secondary roads were closed off, and shelters were opened to house victims of the flood. The *News and Observer* described the scene: "On Sunday, life in Kinston took on a surreal quality. Hordes of disaster-watchers with video cameras on their way to flooded neighborhoods drove past a long line of hurricane victims waiting downtown for emergency welfare benefits. Motorists on U.S. 70 saw Jet Skis surfing beside them on the other side of the guardrail."

Flooding on the Northeast Cape Fear River was equally destructive. Dozens of developments in Duplin and Pender Counties suffered through the same kind of soggy invasion as communities along the Neuse. Their ordeal was made worse when four additional inches of rain fell on the area in the days following Fran. In Burgaw, scores of homes were flooded immediately after the hurricane, and the waters just kept rising. Homeowners near N.C. 53 used canoes and skiffs to retrieve their belongings. Many homes were hit from two directions: trees that fell during the storm tore holes through roofs, allowing heavy rains to soak furnishings; days later, creeping floodwaters from the river finished the job. In some neighborhoods, the water stood eight feet deep. Countless roads in Pender and Duplin Counties disappeared under the murky flood, and portions of N.C. 53 and N.C. 210 were washed out. Remarkably, dozens of motorists attempted to drive anyway, providing a booming business for Green's Wrecker Service in Maple Hill. According to the *Wilmington Star-News*, Green towed in twenty-eight vehicles in just two days, at $150 each. "It's extra if there's a snakebite involved," owner Leslie Green said with a laugh.

Beyond the borders of North Carolina, flash floods also forced evacuations across the Virginia countryside, prompting dozens of dramatic rescues. As Fran plowed through that state's foothills and mountains, it dropped up to fourteen inches of rain on parts of the region. Swiftly rising creeks and rivers forced some families to flee, while others became stranded in mountain hollows and remote areas. Countless residents had to be plucked from their rooftops and their cars by National Guard helicopters, and rescue teams struggled to reach others with boats and military vehicles. The entire city of Elkton was cut off by surrounding floodwaters, and Rockingham and Page Counties in the Shenandoah Valley were also hard-hit, requiring dozens of rescues. The Dan River basin, which includes Pittsylvania and Halifax Counties and the city of Danville, crested eleven feet above flood stage — one foot higher than the previous record set during hurricane Agnes in 1972. Heavy flooding was reported in the South Boston area, and about forty homes were evacuated after water began spilling over a dam on the Moormans River twenty miles northwest of Charlottesville.

Even in eastern Virginia, Fran will be remembered. Wind gusts reached 71 mph in Hampton, and scattered damages were reported. At Virginia Beach, a 50-by-50-foot chunk of the Ramada Plaza oceanfront hotel came tumbling down, exposing insulation and beams between the sixth and ninth floors. Gale-force winds over the Chesapeake Bay pushed waters up the Potomac River, forcing the evacuation of some areas near Washington, D.C. More than three feet of water covered streets in the Old Town district of Alexandria, marking some of that area's worst flooding in many years. Even beyond Virginia, serious floods forced residents out of their homes in West Virginia and Pennsylvania. Like a handful of memorable past hurricanes, Fran left its distinctive mark on the interior portions of several states.

In the hours and days following Fran's trek across North Carolina, Tar Heel residents struggled to recover from the shock of the storm. Most quickly went to work salvaging what they could and beginning the arduous task of cleaning up the mess the hurricane had left behind. Curfews were established in most cities, as officials attempted to maintain order and prevent looting along darkened streets. As with any hurricane, perhaps the most basic necessity sucked away by Fran was available electric power. The Reuters News Service reported that 4.5 million people in the Carolinas and Virginia were without electricity on the day after the storm. Caravans of utility crews poured into the stricken areas almost immediately, some coming from as far away as Florida and Alabama. Carolina Power and Light reported that it recruited about 2,000 additional workers, bringing its total work force to 7,300 — about 2,000 more than the army of workers gathered in South Florida after hurricane Andrew in 1992. Hotels throughout the Triangle were overbooked, and CP&L scrambled to find rooms for all of its crews — some were even taken in by grateful residents. The power outage was accompanied by all sorts of side effects. For example, about 295 of Durham's 300 traffic signals were knocked out, causing the city's accident rate to climb 30 percent after the storm and directly contributing to at least one fatality. In some communities, residents were fortunate enough to have their power restored within hours. But for many, the wait for crews to untangle twisted lines and replace damaged poles and transformers stretched out for many days. As a result, thousands struggled through the cleanup period without benefit of lights, hot showers, televisions, electric stoves, and other comforts of modern living.

A few people with portable generators cranked up modest supplies of power. In some cases, those without electricity borrowed from neighbors who had it, running lengths of extension cord from house to house. In areas where the power was out for more than a day or two, stocks of frozen foods quickly began to deteriorate, so massive backyard barbecues were hastily assembled from Wrightsville Beach to Raleigh and beyond. These impromptu pot-luck feasts brought neighbors together to talk about the storm and to form alliances of survival skill and kindred spirits.

Immediately following the storm, the four commodities most in demand were ice, chain saws, generators, and gasoline. Because most gas pumps require electricity to operate, there were only a few locations equipped to serve the needs of countless thirsty cars. On the morning after Fran passed through, a station on Old Apex Road near Raleigh was one of the few in the Triangle to have working pumps, and motorists waited in line there for up to an hour to fill their cars. The station's manager had sold gas to about 980 people by noon, when the pumps ran out. The traffic was so heavy she couldn't keep up with the sales and resorted to an honor system for patrons buying gas.

In many communities, and especially in the Triangle, hordes of people lined up at building supply centers like Home Depot and Ace Hardware to buy chain saws, generators, and plastic tarps. First-time do-it-yourselfers waited for hours to make their purchases. At one Home Depot in Cary, store owners had to hold a lottery and erect bleachers in the parking lot to manage the anxious crowds. At one point, an announcement came over the store's public address system that another truckload of generators had just arrived, and in less than a minute, a line of one hundred potential buyers had formed. The store later reported that it had sold about four hundred chain saws and four

hundred generators between the morning of September 6 and the following afternoon.

Similar lines snaked around open grocery stores, where eager shoppers had come to buy ice and other necessities. The ice was used primarily by those struggling to save hundreds of dollars worth of steaks, chops, and other items

removed from their powerless freezers. Packaged foods also sold quickly—
anything that didn't require refrigeration or cooking. Word soon spread re-
garding which restaurants and fast-food grills were serving hot meals, and
thousands of families flocked to the few scattered establishments lucky
enough to have power. Of course, relief agencies of all kinds were also hard at
work across the state, providing meals, clothing, and shelter to those in need.
The Red Cross, the Salvation Army, and numerous church-sponsored groups
set up relief stations and shelters to aid untold numbers of victims. Dona-
tions were collected to fund the relief effort, including large corporate gifts to
the Red Cross. Among the contributions were gifts of $25,000 from the U.S.
Tobacco Company and $150,000 from Glaxo-Wellcome.

The difficult physical task of cutting and removing downed trees kept many
families busy to the exclusion of almost all else. Businesses specializing in tree
cutting and home repairs were flooded with requests, and most filled their
week's schedule within hours that first day. Frustrated homeowners called
dozens of companies and could only reach recordings. In desperation, many
storm victims whose homes were covered with oaks took chain saws in their
hands for the first time, buzzing frantically through the timbers that engulfed their property. For days and even weeks, a whining symphony of saws
echoed through virtually every neighborhood in a forty-county stretch along
the hurricane's path. And as the trees, branches, and stumps were cut and
stacked, mountains of storm debris quickly grew along city streets and country roads. Local officials scrambled to deal with the incredible volume of
waste, which required caravans of trucks to haul and specially designated lots
for mulching and burning. Virtually every city and county in the region managed its own "stump dump."

Fran's winds had a devastating effect on Tar Heel forests. The North Carolina Division of Forest Resources estimated that the storm damaged 8.2
million acres of woodland. In some coastal sections, 85 percent of all trees
sustained some type of damage, and that figure exceeded 50 percent in the
Raleigh-Durham area. Among coastal counties, New Hanover and Pender suffered the greatest destruction; timber losses in Pender County alone exceeded
$89 million. In many forested areas, the tremendous amount of dead wood
greatly increased the likelihood of fire. Fallen trees blocked fire lanes and
breaks, limiting the mobility of Forest Service personnel and restricting access for firefighters. The economic impact on the state's forests alone was estimated at $1.3 billion.

Down at the coast, the cleanup was particularly grueling, especially along
the barrier islands so devastated by storm surge. At Topsail Island, two FEMA
search and rescue teams used dogs and sophisticated listening devices to hunt
for at least five people who had been reported missing. (These same teams had
been deployed to search through rubble after the federal office bombing in

Oklahoma City.) One team had to be airlifted to the northernmost end of North Topsail Beach because the area was completely inaccessible after the storm. In Wilmington, more heavy rains fell two days after Fran, and authorities warned residents not to drive because logs cut from downed trees were floating about in the streets. A particularly difficult problem to combat was the influx of gawkers who drove about the hard-hit coastal region with video cameras pressed against their windshields, slowing traffic and obstructing the recovery effort. Even the skies were crowded—so crowded, in fact, that authorities were forced to create a no-fly zone over the beaches where airborne sightseers were making low passes to view the crushed buildings.

Just as they had been after Bertha, hundreds of resort property owners were anxious to return to the affected beaches the day after the storm. This time, however, the destruction was even more severe, and the resolve of town officials was steady. Nevertheless, residents lined up at the bridges, waiting for the first chance to get back on the islands and go to work. Many in the Triangle area were unable to travel to the coast to check on their vacation cottages, so instead they gathered at the state fairgrounds in Raleigh to view a videotape of the destruction. The tape was recorded by the state highway patrol during a helicopter flight down the coast. Trooper Ed Maness, who was behind the camera, told the *News and Observer*, "There were times I had to look up from my view finder and out at the coast to see if all that damage was really real."

Soon after Fran barreled through North Carolina, Governor Jim Hunt declared a state of emergency in all one hundred counties, the first time such action had ever been taken. He was later described as "shocked" by the extent of the damage. Initially, 1,000 National Guard troops were called to action, and the governor ordered 1,400 of the state's prisoners to help with the cleanup. He also announced that state workers living in the affected counties should stay home for up to a week to help their communities recover. Hunt flew by helicopter from the capital to Wilmington, where he rendezvoused with FEMA director James Lee Witt. Together with other officials, they then flew up the coast and toured the battered shoreline from Brunswick County to Topsail Beach. President Clinton initially declared ten counties federal disaster areas, but as the days went by, that number crept upward—by September 11 there were forty, and ultimately more than fifty counties were so designated. On the fifteenth, Clinton made a brief stop in Raleigh, toured hurricane-damaged neighborhoods by helicopter, and met with Hunt and other officials in a live-stock showroom at the state fairgrounds. After pledging to "do whatever we can," he returned to Washington. Ultimately, more than $700 million in federal aid made its way to North Carolina after the storm. In December 1996, the Clinton administration also took the unusual position of agreeing to increase its share of spending for repairs to the state's highways, bridges, and other public works, to 90 percent of the eligible costs.

Estimates for the total economic impact of hurricane Fran vary, depending on which government agency you ask. Calculating these totals for a major storm is an immense task that can sometimes take months to complete. Even then, the variables are many and the estimates often lack precision. In addition to the measurable costs in damaged or destroyed property, there are hidden and indirect costs, such as lost wages, retail losses during power outages, and the lingering impact on tourism along the coast. On the other hand, there are positive impacts on some segments of the economy, such as the construction industry, which typically experiences a boom after major hurricanes.

According to the National Climatic Data Center, Fran caused *at least* $5 billion in damages to property, timber, and agriculture in North Carolina alone. The North Carolina Division of Emergency Management places that total at $5.2 billion. In addition, early estimates from other states affected by the storm were as follows: South Carolina, $40 million; Virginia, $350 million; Maryland, $100 million; Pennsylvania, $80 million; West Virginia, $40 million; and Ohio, $40 million. Therefore, the sum total for Fran approached $6 billion, making it the third most costly hurricane in U.S. history and bumping 1995's Opal into fourth place.

But perhaps the most significant cost was the death of thirty-seven people, directly and indirectly due to the storm. In North Carolina, twenty-four fatalities were reported, including three from drowning, twelve related to fallen trees, and nine due to other storm-related events. Three people died in Onslow County, including Lance Corporal Steve Sears, an Arkansas native who was out driving with two fellow marines the night of the storm. The trio became disoriented in the torrential rains, attempted to drive their Mustang over the high-rise bridge to North Topsail Beach, and were swept into the raging sound. After escaping the sinking car, one of the men made it back to land and one grabbed a tree and was rescued nine hours later, but Sears drowned.

The deaths were geographically distributed along Fran's path. Several victims were killed when large trees crashed through their homes, including teenager Cristina Marie Foust in Onslow County, Donald Davis in Benson, Mary Bland Reaves in Raleigh, and eighth-grader Curtis Wayne Warren in Alamance County. In Duplin County, Rose Hill resident Marion Rouse was killed and her husband injured when their chimney collapsed onto them as they sat in their den. When the couple was discovered some nine hours later by members of the Rose Hill Fire Department, Rouse was found on her couch, which had crashed through the floor. Her husband, though still alive, was pinned under a pile of bricks from the fallen chimney.

Responding to an emergency call at the peak of the storm, nineteen-year-old Bahama volunteer firefighter Ricky Dorsey was another of Fran's tragic victims. He was killed when a falling tree landed on his truck as he was driving down Staggville Road, north of Durham. After the storm, as the cleanup got under way, Zebulon native Walt DeYoung died while attempting to clear debris from his mother's home in Raleigh. DeYoung, a youth hockey coach, was deeply gashed on the legs by a falling tree and later bled to death. A Raleigh teen drowned while swimming in the flood-swollen Crabtree Creek, and a Greensboro man was found floating near the damaged remains of the Scotts Hill Marina in Pender County. He apparently had been helping a friend board up his home when he was swept off a deck and drowned.

One of the storm's more memorable tragedies was that of Georgia Greene. The seventy-five-year-old Surf City woman lived alone in her mobile home and depended on round-the-clock nurses to provide her care. As the evacuation of Topsail Island got under way, Greene refused to leave, stating that she had never left for a storm before. As Fran swept over the island, it demolished her aluminum home, tossing the ninety-pound woman into the raging waters, her mattress beneath her. For hours she drifted through the flooded marsh, enduring the worst of the hurricane's winds and rains. The following day, she was found by rescue workers, still aboard her mattress and still alive. She was taken to a local hospital but died the following day. Cause of death: hypothermia.

Without doubt, Fran's visit to the state established it as the new benchmark for hurricanes in North Carolina—at least for a while. It may not have been the strongest storm to hit the Tar Heel coast in the twentieth century, but its widespread effects and tremendous toll exceeded that of the previous hurricane to set the standard—hurricane Hazel. It had such a significant impact on such a large number of people that a whole new generation of North Carolinians will forever be prepared to reminisce whenever the topic comes up. For decades to come, they'll tell their children and their grandchildren: "I remember Fran!"

BONNIE (AUGUST 26–28, 1998)

At the beginning of the 1998 hurricane season, William Gray issued his annual prediction for tropical activity in the Atlantic Basin, and once again the numbers were up. An active La Niña in the Pacific contributed to the forecast, and emergency planners across the Southeast stood watch for approaching storms. After a relatively slow start, the season cranked up quickly when ten tropical storms formed in just thirty-five days. National Hurricane Center director Jerry Jarrell described the period: "That's nearly a whole year's worth of activity crammed into little more than a month." At one point, on September 25, there were four Atlantic hurricanes in progress at the same time—a rare occurrence that was last recorded in 1893. But the abundance of tropical weather wasn't the only thing memorable about the 1998 season. Before it was

all over, it would become the deadliest Atlantic hurricane season in more than two hundred years.

The real monster of the season was October's hurricane Mitch, a deadly category five that washed away entire villages in Honduras and Nicaragua and claimed over 11,000 lives. At one point, Mitch's sustained winds reached 180 mph and its barometric pressure dropped to 26.73 inches, making it the fourth most intense hurricane ever recorded in the Western Hemisphere. Not since the Great Hurricane of 1780 killed 22,000 across the Caribbean had so

A Jacksonville couple navigates flooded Fountaintown Road during hurricane Bonnie. (Photo by Don Bryan; courtesy of the Jacksonville Daily News)

many lives been swept away by a single storm. Hurricane Georges, the year's other killer cyclone, was near category-five status at its peak on September 20. It went on to batter numerous islands in the northeast Caribbean, taking more than 500 lives well before its eventual landfall near Key West.

Though Georges and Mitch were the deadliest and most destructive of the year's storms, it was hurricane Bonnie that brought eastern North Carolina its third hurricane in three years. For weather watchers along the Tar Heel coast, Bonnie's approach in August provoked an eerie sense of déjà vu. Those who were still repairing the damages from hurricanes Bertha and Fran in 1996 watched the storm's northward swing with a renewed sense of dread. After passing Hispaniola on August 22, Bonnie strengthened to category-three intensity, then slowly drifted northward for twenty-four hours after steering currents collapsed. By the twenty-fifth, the 400-mile-wide storm bore a strong resemblance to Fran — oversized, powerful, and taking dead aim on Brunswick and New Hanover Counties.

Fortunately, Bonnie weakened and slowed as it approached Cape Fear, stalling briefly just after landfall at 5:00 P.M. on Wednesday, August 26. From there the storm's center crept northeast along the coast instead of tracking inland over Wilmington as had been expected. By the time the storm moved over Onslow County early on the morning of the twenty-seventh, it had further weakened to category-two strength. Its slow movement was agonizing to thousands of evacuees who had retreated to emergency shelters to wait out the storm. Slowly its winds dropped, though heavy rains pounded the eastern counties for hours. Bonnie was downgraded to a tropical storm later in the day but then turned to the north and northeast, unexpectedly regaining hurricane strength by 11:00 P.M. As it exited the coast over Currituck County, the revitalized storm surprised many and smacked the Tidewater region with hurricane-force winds and pounding waves. Even local emergency management officials were caught by surprise; they had closed their offices for the night without anticipating that the storm would strengthen. Then, almost as quickly as it had reintensified, Bonnie moved out to sea away from the Virginia shore.

At landfall near Cape Fear, Bonnie's maximum sustained winds were 110 mph, though no instruments in the area recorded winds of that speed. Frying Pan Shoals tower recorded a peak gust of 104 mph, and gusts of 89 mph at Kure Beach and 80 mph at Frisco were also reported. An unofficial reading of a peak gust of 116 mph was taken at Wrightsville Beach. As the storm regained strength while exiting the coast, additional wind extremes were measured. These included gusts of 94 mph at the Currituck County Emergency Operations Center, 104 mph at Cape Henry, Virginia, and 100 mph at Chesapeake Light. According to National Hurricane Center reports, Bonnie's barometric pressure was 28.47 inches at landfall.

Storm tides of five to eight feet above normal were reported along the Brunswick and New Hanover beaches. A storm surge of six feet hit the shores of the Albemarle Sound in Pasquotank and Camden Counties. Water levels of five to six feet above normal were also reported in Washington, Belhaven, and Swan Quarter. As Bonnie moved off the Outer Banks on the night of August 27, strong west winds on the backside of the storm delivered significant soundside flooding to Roanoke and Hatteras Islands.

Even along the coast, localized flooding resulted from the storm's relentless rains. Eight to eleven inches were recorded in several of the thirty counties under the swathe of the hurricane. In Jacksonville, 11.34 inches were recorded during the storm, with 10.4 inches falling in one twenty-four-hour period. Other totals included 11.03 inches at Cherry Point, 9.45 inches at Wilmington, and 8.02 inches in Greenville.

As the storm exited the coast and initial damage estimates got under way, many felt an early sense of relief that Bonnie was not as bad as it could have been. Though significant pockets of destruction were scattered throughout the eastern third of the state, it was immediately evident to local and state officials that this storm was no Fran. Beachfront vacation homes were generally intact; winds and tides were less destructive than had been expected. Still, significant agricultural damages and broadly scattered property destruction were reported from Myrtle Beach to the Virginia line.

In Calabash, several homes were damaged by high winds, and one was reportedly "blown off its foundation." In Shallotte, where several local residents weathered the storm at the Comfort Inn, power was lost and no one had access to television or radio. Hungry for information about the storm, one man used his cell phone to call his father in New York for information via the Weather Channel.

A bank in Sunset Beach lost its roof, as did portions of the Brunswick Community Hospital in Supply. National Guard troops and health care workers from New Hanover Regional Medical Center journeyed to the damaged hospital through torrential rains at 1:00 A.M. to transfer several patients to safe quarters in Wilmington. Along the way, their convoy of Humvees was nearly swamped by deep standing water that covered the highway. It was then that they apparently picked up an unexpected passenger. Robyn Moore, a mental health director from Wilmington, told the Associated Press: "When we hit the puddle, water went over the top of the car, and we heard this huge THUMP. . . . We see this thing on the windshield, this big grayish-brown thing." When they arrived at their destination, they found a dead water moccasin entangled in the Humvee's front grille.

As the storm's fifteen-mile-wide eye slowly passed near Southport, many residents ventured out into the relative calm. Four golfers even risked playing a few holes at nearby St. James Plantation resort while the winds abated. For the most part, damages to the Brunswick beaches and inland communities were not significant. Signs and tree limbs were downed and scattered, and the occasional power pole was twisted by Bonnie's winds. Early property damage estimates in Brunswick County were around $20 million.

As with many hurricanes, rumor control was a problem during the storm. Radio stations falsely reported that both Holden Beach and Caswell Beach were severed by surging tides that created new inlets on the barrier islands. Neither report was true, although Caswell Beach was temporarily closed when waves washed over the roadway. Another rumor suggested that forty-six Bald Head Island residents were trapped inside the island lighthouse when Bonnie struck. Actually, six residents, four dogs, two cats, and a parrot did spend the night in the historic light, but they stayed there by choice and were all safe. This story made its way to CNN's *Headline News* before local officials could correct the facts.

In New Hanover County, more than a dozen homes were heavily damaged, including several in Carolina Beach. Wrightsville Beach fared somewhat better, with only minor structural damages reported. But given the large size and apparent strength of the storm as it made landfall, the damages were less than had been expected in the resort communities that face the Atlantic. The lack of significant destruction along the oceanfront here was credited to two factors: a moderate storm surge and recently completed sand nourishment projects.

Storm surge heights of nine to twelve feet had been forecast on the New

Hanover beaches, but the actual surge level was around eight to nine feet. According to the *Wilmington Star-News*, water level marks at Wrightsville Beach were four feet below those measured after hurricane Fran. Costly sand nourishment that became a priority after Fran had fortified much of New Hanover's beaches with new protective dunes and stabilizing grasses. In general these dunes held up well against Bonnie, protecting rows of cottages and island streets from the storm's battering waves. But in some locations, the surging ocean washed right through the manmade barriers. Wrightsville Beach saw overwash in only a couple of spots, but on the northern end of Carolina Beach, deep sand filled streets and swallowed mailboxes, just as it had during Fran.

Fishing piers are always vulnerable to hurricanes that sweep the coast, and at least seven in North Carolina were shaved down by Bonnie's rolling surf. Among them was Freddy Phelps's pier on the northern tip of Carolina Beach. Bonnie was the fifth hurricane to whittle away at this structure since it was first built in 1979, though on Phelps's pier-destruction scale, it was also the weakest. According to the *Star-News*, Bonnie lopped off 45 feet of the pier's length, compared with 500 feet in hurricane Diana (1984), 400 feet in Hugo (1989), 200 feet in Bertha (July 1996), and the entire 850-foot structure during Fran two months later. "I got tired of looking at it during the storm," Phelps told the *Star-News* following Bonnie. "I mean, we just finished it. I had to take a seat on the couch for a while."

Up the coast at Topsail Island, the damages were serious in places, but not nearly as extensive as those left by Bertha and Fran in 1996. While Surf City's rebuilt dunes were breached in some spots, they held together in most locations. At least five trailers at Rogers Bay Campground were found floating in the sound, which engulfed a good portion of the recreation area near Island Drive. Topsail Beach suffered around $1.8 million in damages. At North Topsail, many dunes were leveled, creating waves of sand that covered S.R. 1568 and washed away a twenty-foot section of the roadbed. Sixty percent of the twelve-mile dune wall erected following Fran was devoured by Bonnie's waves. Garage doors, heat pumps, decks, and roofs were half-buried along the roadway, but the town was in far better shape than it had been two years earlier. In all, about six hundred homes suffered minor damages, ten buildings were condemned, and total damages were estimated at $2.5 million.

Carteret County had scattered losses, most of them concentrated in the beachfront portions of Emerald Isle. Several homes lost their roofs, and others had septic tanks uncovered or washed away. A 150-foot section was ripped from the Iron Steamer Pier in Pine Knoll Shores, and the Indian Beach Pier lost 100 feet. An apparent waterspout struck a dormitory at Duke University's marine lab near Beaufort, stripping away the building's roof and causing its walls to collapse. All students had left the site before the storm struck, and there were no reported injuries. In Morehead City, Beaufort, and Emerald Isle, at least seventy-two structures sustained major damage.

From Fort Macon to Emerald Isle, residents emerged after the storm to find a junkyard mess on the beach—thousands of automobile tires were strewn about in the surf and on the sand as far as the eye could see. They had broken free from old artificial reefs deposited about three miles offshore by the state's Marine Fisheries division. The tire-reefs, which had been strung together with nylon and steel cables, were originally placed to attract small fish, which attract big fish, which in turn attract fishermen. Some 100,000 tires were sunk offshore back in the 1970s along with an old jet, two barges, and a decommissioned ship. Some of these tires broke free under the strain of Bonnie's churning waters and were deposited on the Carteret coast. The Fisheries Division, which no longer uses tires for reefs, responded quickly to the scene. Along with a hundred prisoners and a National Guard unit, Division personnel worked to clear the tires from the beach.

All along Bonnie's course over eastern North Carolina, isolated tornadoes caused significant destruction. Several touched down as the first rain bands of the storm came onshore during the afternoon and evening of August 26. Some damages were confirmed to have resulted from these tornadoes, while others were judged the likely result of straight-line winds. A tornado moved through the Riverdale area of southern Craven County, lifting one house off its foundation and damaging at least eight others. A Duplin County fire station was knocked off its foundation by another twister, and buildings in Holly Ridge were peeled open by yet another. Oriental and Edenton also had confirmed tornadoes. Other locations hit by suspected tornadoes included five homes and another fire station near Pink Hill, several storage buildings in Tyrrell County, and homes in Maysville, Wenona, and Gum Neck.

Wind-driven floodwaters ripped away docks and inundated homes in numerous coastal towns. Water levels five to six feet above normal were reported in Washington and Swan Quarter. Manteo and sections of Highway 12 on Hatteras Island were flooded with water levels two to five feet above normal. High waters flushed some forms of wildlife from their usual habitats, depositing hundreds of minnows in the front yard of one Washington Park residence. A Stumpy Point man killed eight snakes with a rifle as he waded along a road near his home.

But of all the river towns that dealt with Bonnie's rising tides, none was flooded as badly as Belhaven. Founded on the Pungo River by a lumber magnate in the 1800s, this town of 2,400 lies entirely in the hundred-year floodplain, a mere three feet above sea level. Its streets were buried under four feet of water during Bonnie's visit, turning the entire town into a down-east version of Venice, Italy. This wasn't the first time Belhaven had flooded. In fact, the town had been swamped three times in the previous two years by two hurricanes and a tropical storm. But residents were especially frustrated with Bonnie. Work had just started on a $9.2 million federal project to elevate more than 370 homes, though only three were under way when the storm struck. As a result, furniture and household goods floated through the streets along with docks, soft drink machines, and propane tanks. One resident said the current was so strong that he had to walk sideways to keep from being knocked down. In the end, about 40 percent of the town's 2,000 homes and two-thirds of its sixty businesses were damaged by high water.

Before moving out to sea, Bonnie slowed to a crawl over the tidewater region, unleashing fierce northeasterly winds and heavy rains onto Virginia Beach and across the northeastern counties in North Carolina. Along Atlantic Avenue, where Virginia Beach's resort hotels face the ocean, shattered glass and twisted awnings lined the street. Through the night, small trees and power lines were downed, all from a storm that had been downgraded and was supposed to be dissipating. More than 300,000 Virginia Power customers were without electricity during the night. Bonnie's reformation to hurricane status caught the region's local officials off guard; here the damages were greater than had been expected.

Among the storm's few tragedies was the death of a twelve-year-old Currituck County girl who was killed by a falling tree. Barbara Caley, a sixth-grader at Central Elementary in Barco, died when an eighty-foot-tall sycamore tree struck the Caley home during Bonnie's resurgence on the night of August 27. A few other fatalities were indirectly related to the storm, including a Myrtle Beach man who was electrocuted while checking his home generator, a rip-current victim in Rehoboth Beach, Delaware, and a man who drowned when his rowboat capsized near Cape Cod, Massachusetts. A New Jersey man was among those believed to have drowned when a two-day search by divers and helicopters failed to yield a body. Antonio Mandarino was reportedly carried under by Bonnie's pounding surf while swimming at Point Pleasant Beach. But Mandarino later turned up alive and well and in police custody. He and his fiancée were arrested for trying to fake his death to avoid bad check charges in Bergen County.

In the days following Bonnie, media reports quoted state officials as estimating total damages in North Carolina at more than $1 billion, or "about the same as Bertha." A good portion of the losses was attributed to crop damage, which was particularly significant to tobacco, corn, cotton, and soybeans. Total agricultural damages in the state were around $164 million, with Sampson, Columbus, and Jones Counties being hit the hardest. Governor Jim Hunt, after touring the coastal counties, described the scene for the Associated Press: "You fly along and don't see much damage to the beach houses, and it's easy to think we didn't have much damage. . . . But then you look at the tobacco in Don Sweeting's fields and you know the damage has been extensive." Thirty counties in the eastern third of the state were declared federal disaster areas, and more than 37,000 property insurance damage claims were filed in the first weeks after the storm—about twice the number filed following Bertha.

A more conservative damage calculation comes from the National Hurricane Center, which routinely takes insured property claim figures from the American Insurance Services Group and doubles them to achieve total storm damage estimates. By this reckoning, hurricane Bonnie caused about $480 million in damages in North Carolina and around $720 million in the U.S. as a whole. Carolina Power and Light Company, which spends millions for additional manpower and equipment to restore electricity after hurricanes, has its own scale for measuring each storm's impact. Bonnie cost the company $25 million, making it the second most expensive hurricane in the utility's history to that date. Fran cost a record $95 million, while Bertha's tab was about $11 million and Hugo's was $12 million.

With the growth of the Internet in the late 1990s, more and more people began turning to their personal computers for online information about approaching hurricanes. Bonnie was no exception, as websites that featured forecast positions, satellite and radar images, and other timely information were flooded with hits. Those residents who did not lose power and who stayed up with the storm were able to watch its progress through the state in the form of bright green swirls across their computer monitors. And major websites like that of the National Hurricane Center saw unprecedented activity. To meet the heavy demand, the Federal Emergency Management Agency's (FEMA) popular website was upgraded even as Bonnie was passing over the North Carolina coast. A new fiber optic cable was installed on the evening of the twenty-sixth, expanding the site's bandwidth tenfold. During the storm's passing, the FEMA site was deluged with more than two million hits each day.

While Bonnie was spinning over eastern North Carolina, more than a dozen families from Greenville watched the storm's progress from hotel rooms in Williamsport, Pennsylvania. They were there to support the Greenville All-Stars, a youth baseball team taking part in the Little League World Series. According to the *News and Observer*, before their game against a California team the parents and fans from Greenville were heard chanting from the stands:

Hey, Bonnie go back to the ocean;
Hey, Bonnie go back to the sea;
Hey, Bonnie go back to the ocean;
And get out of Greenville, N.C.

Surrounded by reporters, Governor Jim Hunt inspects damaged tobacco fields in Onslow County following hurricane Bonnie. Storm-related agricultural losses through the late 1990s amounted to more than $1.5 billion. (Photo by Don Bryan; courtesy of the Jacksonville Daily News)

A determined beachgoer relaxes among thousands of automobile tires scattered along the strand in Pine Knoll Shores. The tires, once part of a large offshore artificial reef, were freed by strong currents and waves during hurricane Bonnie. (Photo by Barbara Mather; courtesy of the Jacksonville Daily News)

DENNIS (AUGUST 30–SEPTEMBER 5, 1999)

Through the early weeks of summer 1999, tourism was booming for most of the resort properties and beach businesses that cater to seasonal crowds on the North Carolina coast. Even though another dire hurricane forecast had been issued for the Atlantic, vacationers packed the beaches to soak up sun and play in the surf. The peak tourist season was becoming increasingly compact, however, as families planned their vacations around return-to-school schedules that crept toward early August. But property agents noted that those interested in beach house rentals were beginning to avoid late August bookings for another reason—the potential threat of yet another hurricane on the Tar Heel coast.

As residents and vacationers looked ahead to the approaching Labor Day holiday, their attention was soon focused on hurricane Dennis, a larger-than-average storm that formed near the Bahamas on August 24 and reached hurricane strength two days later. It produced near-hurricane conditions at Abaco Island on the twenty-eighth and then intensified later that night, building sustained winds of 105 mph. It maintained category-two intensity for the next two days while paralleling the southeastern U.S. coast about 115 miles offshore. As it began its approach to the Carolinas it slowly weakened but moved to within about sixty miles of that frequent hurricane target, Cape Fear. For a while, it looked like the beaches from Wrightsville to Topsail would catch another direct hit.

Throughout its development and movement northward, Dennis was never a classic tightly wound hurricane. Westerly wind shear persisted throughout much of its course, and its large, forty-mile-wide eye was often poorly defined. Once the storm closed in on North Carolina on August 30, it encountered a midlatitude trough that steered it toward the north and northeast, pushing it out to sea off Cape Hatteras. At the same time, extensive weakening occurred. This was good news at first, since the southeastern portions of the state were spared landfall. But Dennis's large wind field extended for two hundred miles in every direction around the storm's center, pushing tropical-storm-force winds of 45 to 65 mph onto the beaches.

As the trough passed and moved out to sea on the thirty-first, all steering currents collapsed and Dennis began its wayward drift about 125 miles east-southeast of Cape Hatteras. It was about this time that the hurricane weakened further, as westerly shear returned and cool, dry air became entrained in the circulation. On September 1 it was downgraded to a tropical storm, though forecasters at the Hurricane Center said by this time it may have been as much a subtropical or extratropical cyclone as a tropical cyclone. Still, as Dennis drifted aimlessly over the offshore waters of the Outer Banks, storm-force winds drove waves and water along the shore from Cape Lookout to New Jer-

sey. The waves battered the beaches for days, and steady northeast winds pushed water up the rivers and sounds. The rain fell steadily, too, and many along the coast suffered through the storm that wouldn't go away—the storm the press appropriately dubbed Dennis the Menace. Steve Lyons, the Weather Channel's senior meteorologist, told the *Virginian-Pilot*: "It is meandering out there. It's a tumbleweed, heading back from whence it came. But this is going to be a long, drawn-out ordeal. It's going to be here for another few days."

After drifting erratically for two days, the storm was finally edged southward by a large westerly ridge on the evening of September 2. This front directed it over slightly warmer water, where a burst of convection was observed on the following day. The unpredictable storm began to intensify again and did another about-face—this time turning northwest, directly toward Cape Lookout on the North Carolina coast. The reintensification continued

Corrie Stevens, a vacationer from Philadelphia, waits for a tow truck after driving his Cadillac into a canal on the Old Oregon Inlet Road in South Nags Head. Hurricane Dennis disrupted vacations for thousands along the North Carolina coast. (Photo by Drew Wilson; courtesy of the Virginian-Pilot)

throughout its last dash for land, though it never again reached hurricane strength. Dennis finally moved ashore over the Cape Lookout National Seashore just before 5:00 P.M. on Saturday, September 4. From there it tracked inland, weakening to a depression as it dissipated over central North Carolina on the following day.

Dennis was not considered a hurricane at landfall, but its first passage near the North Carolina coast on August 30 may have produced sustained winds of near-hurricane force along portions of New Hanover, Carteret, and Dare Counties. At Frying Pan Shoals, automated equipment recorded sustained winds of 94 mph, gusts to 112 mph, and a barometric low of 28.86 inches. Other high gusts were measured elsewhere along the coast during this period, including 111 mph at Wrightsville Beach, 98 mph at Hatteras Village, 91 mph at Cape Lookout, 89 mph at Oregon Inlet, and 87 mph at Harkers Island Bridge. After losing strength for several days and then reintensifying before its eventual landfall on September 4, Dennis came ashore just below hurricane strength. Hurricane Hunter aircraft reported that it had sustained winds of around 70 mph and a pressure of 29.06 inches at the time of landfall.

Dennis affected the Carolina region twice within a week, even as other weather systems passed through the area during the same period. According to National Weather Service officials, this made determination of the storm's total rainfall somewhat difficult. The heaviest accumulation occurred at Ocracoke, where 19.13 inches were recorded during the week. Southport received 13.50 inches, and some portions of eastern North Carolina reported 6 to 10 inches. Other locations in the eastern third of the state generally reported 3 to 6 inches. Although this rainfall provided a welcomed break from the drought conditions that had gripped the region throughout the summer, it also filled rivers and creeks and set the stage for the disastrous floods of hurricane Floyd two weeks later.

As Dennis approached, stalled, and meandered about, it produced heavy surf, severe erosion, and coastal flooding from Brunswick County to eastern Virginia. Storm surges on the ocean beaches were greatest on August 30 when hurricane Dennis first passed, and on September 4 when tropical storm Dennis moved over Carteret County. Tides generally ranged from three to five feet above normal. But the storm's continuous northeast winds brought prolonged flooding to the state's rivers and sounds. These winds pushed water up onto the north-facing shorelines of numerous riverfront communities. Areas along the Neuse River reported tides of eight to ten feet above normal on August 30, and portions of the Pamlico River saw similar flooding on September 4. Residents of Cedar Island, Ocracoke, and Pamlico County reported that these were the highest water levels they had seen since the September storm of 1933. In Craven and Beaufort Counties, officials reported that several homes were filled with water for the first time since they were built. Beaufort

County suffered more flooding damage from Dennis than from hurricanes Bertha, Fran, or Bonnie.

Though it never made landfall as a hurricane, Dennis's impact on North Carolina was still significant. In addition to the widespread flooding of the state's sounds and eastern rivers, the Outer Banks were particularly hard hit, hammered by high winds and waves that endured for more than 140 hours. The beach towns of Hatteras, Rodanthe, Nags Head, and Kitty Hawk were pounded through almost a dozen lunar high tides, periods when heavy surf carved away sand and overwashed streets and yards. Severe erosion resulted, undermining oceanfront structures and severing the main artery on Hatteras Island—the tenuous pavement of Highway 12. Fifteen-foot-high waves broke steadily along the strand, devouring from thirty to two hundred feet of precious Outer Banks beach sand.

As the storm battered the shore, researchers at the U.S. Army Corps of Engineers's field research pier in Duck took instrument readings and digital photos of the monstrous waves that rolled by. From the end of their concrete pier, which extends almost a quarter-mile into the ocean, they recorded waves of 20 feet, 6 inches—the third-highest swells measured at the site since it first opened in 1977. The highest were 22-foot, 7-inch waves during hurricane Gloria in 1985, followed by 21-foot, 4-inch waves measured in hurricane Gordon in 1994. The engineers noted that waves to the south off Hatteras Island were likely much higher. The U.S. Geological Survey was also on hand to study Dennis's effects and to measure and map the Outer Banks after the storm. According to their report: "Sea level, including tide and storm surge, was near record level. Preliminary calculations by the USGS of the elevation of wave runup on the beach (that is, the highest reach of the waves) was among the highest (.05%) on record."

Very near where Dennis made landfall on Carteret County's Core Banks, a new inlet was created that severed the narrow strip of beach. The inlet, some three hundred yards wide and nearly twelve feet deep, formed just two miles north of one dredged in 1997 by the Corps of Engineers at a cost of $1.4 million. Though inlets provide essential access for commercial fisheries, neither of the Core Banks channels were easy to navigate after the storm. Both were lined with shallow shoals and sandbars that shifted from day to day. The new inlet received little attention in the days following Dennis because of its remote location on an uninhabited island in the Cape Lookout National Seashore.

Further north, the well-populated villages of Hatteras Island were also divided by the rising Atlantic. Just north of Buxton, an inlet was cut across the island, washing away a 3,000-foot section of Highway 12 near Canadian Hole, an area often prone to ocean overwash. A convoy of three N.C. National Guard vehicles discovered the breach on September 1 as Guard members attempted to deliver fuel, water, and food to nearly 5,000 people who failed to evacuate the isolated south end of the island. They reported that the channel was eight feet deep in some locations, too deep for their trucks to cross. Along the way, the Guard saw parts of the roadway that were buried under four feet of sand. The washout severed phone lines on the island, cutting communications between the northern and southern ends. The convoy eventually turned back and delivered 3,700 meals to three fire stations north of Avon. Later that day, National Guard helicopters ferried food, water, and emergency workers into Buxton and transported a pregnant woman out. The storm forced the postponement of the scheduled relighting ceremony at the Cape Hatteras Lighthouse, but the lighthouse weathered the storm without a problem, having been recently relocated to its new site farther away from the Atlantic.

Other portions of the Outer Banks also did battle with Dennis. On Ocracoke Island, nearly seven hundred residents chose not to evacuate and remained through the storm. Electricity was lost early in the week, and the entire island was powered by generator for several days. More than a foot of water was said to have stood in the streets of Ocracoke Village. Along the northern stretches of Hatteras Island, several vehicles were found buried in sand up to their door handles. Even a wrecker was abandoned by its driver after being caught in a deep drift. Beachfront erosion was heavy in Rodanthe, where waves washed over the swimming pool of an oceanfront motel and claimed at least five houses. These houses had already been condemned, but after the storm one resident described them as "physically missing."

At Kitty Hawk, a woman apparently was blown off her deck but not seriously injured. The Kitty Hawk Elementary School lost a portion of its roof to high winds, and at least forty nearby homes were damaged. Most of the damage in Nags Head was on the south end, on Old Oregon Inlet Road. Among the nearly two hundred damaged homes, seven had major damage and one was destroyed. At Corolla, a National Guard Humvee led a caravan of eleven families down the beach at low tide to escape the rising water. One family had to wait for the next low tide.

"You just don't believe it," Debbie Bell told the *News and Observer* as she cleared lumber and debris from the front yard of her surf shop in Rodanthe. "Third-row houses are now oceanfront. The first row is gone. There's always erosion. But how much there was, it's just unreal. I never expected to see that much."

In Carteret County, the hardest-hit areas were at the county's eastern end, where strong northeast winds pushed Pamlico and Core Sound waters onto land. Cedar Island residents suffered through severe flooding on the storms's first pass, then endured more high water when landfall occurred five days later. They referred to the two events as Dennis One and Dennis Two. Those who had lived in the close-knit community long enough described the flooding as the worst since 1933. One resident even reported that the water in his front yard was "up to the top of the steering wheel" of his car. Most agreed that the storm's first pass brought the greatest flooding.

Other nearby communities that saw extreme high water on August 30 included South River, North River, Davis, Stacy, Sea Level, and Atlantic in Carteret County and Harlow, Woodson, Bridgeton, and Sandy Point in Craven County. Many of the flood victims who had spent days cleaning their homes of muck after the storm's first visit could hardly handle the grief that came with round two. Fishing equipment was damaged, cemeteries were flooded, septic systems were swamped, and homes were filled with mud and debris. Among the losses for one South River resident was a prized collection of three classic Mustangs. In all, Dennis damaged more than 26 area businesses, heavily damaged more than 72 homes, and left 230 with minor damage. One local resident told the *Carteret County News-Times*: "You know it's getting bad when Jim Cantori (reporter for the Weather Channel) is in Atlantic Beach."

Across the sound in Pamlico County, the flooding was described as "the worst in memory." Unlike Carteret County, which endured the highest water during the storm's first pass, residents throughout Pamlico County experienced the greatest flood on September 4 as tropical storm Dennis moved inland. Water flowed three feet deep over portions of N.C. 304 and N.C. 55, isolating the communities of Oriental and Hobucken. Beaufort and Hyde Counties also faced extreme tides during the second assault. Low-lying areas in Washington were inundated by the rising Pamlico River, causing city streets to flood and stalling cars. According to reports from the Hurricane Center, flooding along the river was eight to ten feet above normal.

Belhaven was swamped by Dennis, just as it had been through Bertha, Fran, and Bonnie. The low-lying town was turned into an island by the rising Pungo River, which flooded homes and businesses in familiar fashion. As the waters rose, officials moved the town's police operations from their normal quarters to the municipal building on higher ground. After the storm, town residents went to work just as they had so many times before—using garden hoses and brooms to wash away a dark layer of filth that seemed to cover everything. Because of the regular flooding in recent years, the *News and Observer* reported that some Belhaven citizens had "started measuring hurricane high-water marks like some people measure the growth of their children." A pole in Bud O'Neal's drugstore was labeled for comparison. "This is Bertha and here's Dennis," O'Neal told reporters. "Not as bad as Fran. Bad enough though."

Throughout the mid-Atlantic states, six deaths were directly or indirectly attributable to Dennis, including four surf-related drownings in Florida and two fatalities from an auto accident in North Carolina. The accident was a head-on collision that happened near Richlands as 60 mph gusts and two inches of rain pummeled the region. Injuries were widely reported during the six-day storm, especially in those communities that suffered the greatest flooding. And an F2 tornado associated with Dennis touched down near Hampton, Virginia, injuring fifteen people, six seriously.

Though less destructive overall than other recent Tar Heel storms, hurricane Dennis (and then tropical storm Dennis) will long be remembered in many of North Carolina's coastal communities. Crop damages were significant, coastal erosion on the Outer Banks was the worst experienced in years, and hundreds of families dealt with the exhausting cleanup of their homes. Dennis wore people down simply by lingering for so long, but it also left behind significant property damages. On the Outer Banks alone, more than 1,600 properties were damaged, with 5 percent of those destroyed or uninhabitable and 20 percent suffering major damage. The Hurricane Center's conservative estimate was $157 million in agricultural and property damages in North Carolina and Virginia. In the end, though, the storm's impact on eastern North Carolina would be overshadowed by the greater disaster that followed two weeks later—the record-breaking floods of hurricane Floyd.

FLOYD (SEPTEMBER 16, 1999)

There were thirteen people on the roof, that cold stormy night. It was dark, and every sound you would hear would scare you. My dad, my uncle, and my grandpa had to wade through the water to get to the boat. My dad got gook and gunk all over him when he got back. The water had made a hole in the ground, my uncle fell in it. It scared all of us. We thought he was a goner.
—Abby McDonald, from Spooky Waters, *a 1999 essay for her seventh-grade class at Jones Middle School in Trenton*

In the days following the dissipation of tropical storm Dennis, the skies over eastern North Carolina returned to their normal pattern for late summer—hot, humid, and peppered with frequent afternoon thunderstorms. Rivers, streams, and ditches across the eastern counties flowed heavily with water from Dennis's rains. Saturated soils took in what they could, but gave up the excess to flow downstream. More showers fell through early September, with some counties receiving as much as ten additional inches over the course of one week. These rains refilled creeks and swales and set the stage for the arrival of a more ominous rainmaker—and monstrous flood producer—that would later be recognized as the greatest disaster in North Carolina history.

Hurricane Floyd emerged from the central Atlantic at the peak of the busy 1999 season and developed into the year's most awesome hurricane spectacle. It was indeed a very busy year, amidst an extremely active period of tropical activity. The 1999 season produced twelve named storms and eight hurricanes, five of which were major hurricanes. In fact, the five major storms that year—

A toddler looks back on a demolished Oak Island cottage, one of forty destroyed in that community by Floyd's storm surge. In addition, Floyd heavily damaged almost three hundred homes and scuttled portions of two local fishing piers. Not since hurricane Hazel in 1954 had Oak Island suffered such significant property losses. (Photo by Jim Harper; courtesy of the State Port Pilot)

Though heavy rains were already pouring over North Carolina on September 15, hurricane Floyd was still far from landfall. (Photo courtesy of NOAA and Johns Hopkins University)

Bret, Cindy, Floyd, Gert, and Lenny—all reached category-four strength, a phenomenon never before observed since reliable record keeping began in 1886. And the busy 1999 season capped off a dizzying five-year stretch that produced more Atlantic hurricanes than any similar period in recorded history. From 1995 through 1999, forecasters at the National Hurricane Center worked around the clock to keep up with sixty-five named storms, forty-one hurricanes, and twenty major hurricanes with winds exceeding 110 mph. A number of these storms eventually affected North Carolina, including hurricane Fran, which established itself as a new standard for Tar Heel hurricanes when it rolled across the state in 1996. But records were made to be broken, and it didn't take long—just three years—for hurricane Floyd to replace Fran as the state's preeminent weather event.

Floyd's course to America can be traced back to a tropical wave that formed off western Africa on September 2. Growing steadily in size and strength, from a depression on the seventh to a named tropical storm by the eighth, Floyd became a large hurricane at 8:00 A.M. on September 10. Its westward course steered it well above the vulnerable islands of the eastern Caribbean

and far enough north to also avoid Puerto Rico and Hispaniola. While gathering steam on its approach to the eastern Bahamas, Floyd amassed a considerable volume of tropical moisture and grew to become a large, dynamic hurricane. Little or no wind shear and a ridge of high pressure aloft provided the setting for the storm's intensification. Satellite photos told the story as this monster reached its meteorological peak on September 13. With maximum sustained winds of 155 mph and a barometric low pressure of 27.20 inches, Floyd was a borderline category five when it moved to within just three hundred miles of the central Bahamas.

All eyes in the eastern United States were fixed on this large and dangerous cyclone, whose own eye was as pronounced as that of any classic hurricane. The storm's size was immediately impressive too: televised reports in Florida compared satellite images of Floyd with pictures of hurricane Andrew from 1992. Though the two storms reached a similar intensity at nearly the same position in the Atlantic, Floyd's diameter was about three times what Andrew's had been, covering an impressive swathe of the western Atlantic. Hurricane-force winds extended more than 125 miles out from the eye, and the overall storm measured more than 400 miles across. Its formidable size and strength commanded the full attention of everyone with a television or radio from the eastern Bahamas to Florida and the southeastern U.S. coast.

Bahamians were the first to be blasted by the storm. On the night of September 13, Floyd changed heading and turned toward the west-northwest,

passing only about thirty miles north of San Salvador and Cats Island. Fortunately, some weakening occurred during the evening, for on the morning of the fourteenth the storm's eyewall passed directly over central and northern Eleuthera. After a continued turn toward the northwest and more fluctuations in intensity later that afternoon, Floyd struck Abaco Island with at least category-three wind speeds. Few news reports filtered out of the islands during the storm, but it was later learned that many beachfront properties were devastated. At least one Bahamian was killed, and damages were widespread throughout the affected islands.

All through this period, emergency managers and residents on Florida's east coast were anticipating the worst and scrambling to execute their plans. A hurricane watch was issued for portions of the Florida coast on the morning of the thirteenth; this watch was shifted northward and upgraded to a warning later that day. The storm appeared to be lunging toward the Cape Canaveral area, and most people on the eastern side of Florida packed their cars and fled. Almost seventy shelters opened throughout the coastal counties, and another seventy were put on standby. From Miami-Dade to Jacksonville, a mass exodus was under way as 1.3 million people attempted to flee the path of the storm. Highways were jammed with cars, NASA's Kennedy Space

Center was shut down, and for the first time in its twenty-five-year history, Walt Disney World in Orlando closed.

The hurricane continued to spin toward the northwest, but as the hours rolled by it became apparent that Florida would be spared a direct hit. Hurricane warnings were extended northward as the storm's center tracked parallel to the coast. Though Florida's coastal counties were spared the brunt of the worst winds, high surf crumbled piers and caused significant erosion along the beaches. The highest reported sustained winds were 55 mph, with some gusts reaching 75 mph. As the eye passed 110 miles east of Cape Canaveral early on September 15, a data buoy rode mountainous seas 120 miles offshore, measuring fifty-five-foot waves every seventeen seconds. Floyd was still large and dangerous, and now it was edging up the coast toward the Carolinas.

Massive evacuations continued around the clock in the coastal regions of Georgia, South Carolina, and North Carolina. As Floyd neared the lowcountry of South Carolina, many residents near Charleston fled under the stress of great emotion. They still held vivid memories of hurricane Hugo, a similar category-four storm that had struck the region almost exactly ten years earlier. Once again, highways were packed with cars, trucks, and campers that crept inland at a pedestrian pace. Along one portion of Interstate 26, motorists traveled just fifteen miles in four hours. Some even left their cars to engage in fistfights on the highway median. Evening news broadcasts focused on the approaching storm but highlighted the frustrations of roadlocked evacuees who spent long hours trying to escape the freeways and find available hotel rooms.

Evacuations on the North Carolina coast were just as thorough, though they involved smaller populations and fewer headaches. In all, the hurried inland flight of millions of residents, from South Florida to the Virginia coast, was the largest and most complete on record. Some estimates suggested that almost three million people along the southeastern seaboard left their homes and moved to safer quarters in advance of the storm. Emergency planners and government officials were awed by the magnitude of the evacuation. Joe Myers, Florida's emergency management director, told the *Miami Herald*: "In the truest sense, this probably has become the largest peacetime evacuation in U.S. history, and certainly in the state of Florida. This is the first time we've ever had a complete evacuation on either coast."

While forecasters watched the storm hour by hour, residents of coastal North Carolina listened to each news update with a familiar sense of dread. Though Floyd had weakened from its strong category-four peak, it was still a dangerous hurricane as it passed by the Georgia coast. And its future course was now of greater concern to the Tar Heel state. Storm-weary inhabitants from Brunswick County to the Outer Banks were accustomed to plotting hurricanes with curving tracks; they knew that the more Floyd edged toward the west, the less likely it became that it would curve harmlessly out to sea. Fortunately, though, the storm slowly weakened throughout the day on September 15, with sustained winds dropping from 140 mph to 110 mph. But Floyd was so large that even as its breezes still fanned parts of South Florida, its outer rain bands were spreading over New York state. It was during this time, while the storm was still a full day away from landfall, that drenching rains began to fall across eastern North Carolina.

About the time Floyd made its turn just off the Florida coast, emergency managers and local meteorologists began to issue dire warnings for the Carolina beaches—or wherever landfall might eventually occur. Storm surges in excess of fourteen feet were discussed, and residents along the state's barrier islands prepared for the worst. But the continued drop in the hurricane's intensity throughout September 15 was good news for those who owned property at the coast. With Floyd's weakening trend and its downward transition to a category-two storm, a sense of relief suffused hurricane watchers everywhere. Though still oversized when viewed from space, the once-mighty cyclone no longer packed the punch of a killer—or so it seemed.

Floyd finally made landfall around 3:00 A.M. on September 16 at Cape Fear, a point of land that often seems to draw hurricanes like a magnet, having been the scene of recent landfalls during Bertha, Fran, and Bonnie. As Floyd approached the coast, it lost much of its eyewall structure and became less well organized. Reconnaissance aircraft recorded maximum sustained winds of 105 mph at the time of landfall, though no reporting stations experienced sustained winds of that speed. The storm's center dragged along the New Hanover shoreline and then moved inland over Pender and Onslow Counties. Later that day, it passed over New Bern, Washington, and extreme southeastern Virginia, where greater Norfolk was also pounded by heavy rains. Floyd then weakened to a tropical storm and accelerated toward the north-northeast, skirting along the Delmarva Peninsula and up the New Jersey coast. During this time its forward speed increased to 34 mph. As it continued northward early on September 17, it took on extratropical characteristics while buffeting Long Island and coastal New England with high surf and heavy rains. Within two days Floyd became part of a large extratropical low over the North Atlantic and was no longer a distinct storm.

At first look, Floyd was not a meteorological marvel when it came ashore in eastern North Carolina. The storm's loss of strength before landfall was a blessing to those who had watched it teeter near category-five territory just days before. Winds gusted above 100 mph in numerous locations, but wind-related damages were not significant. Still, at the New Hanover County Emergency Operations Center a peak gust of 130 mph was reported, and an instrument atop the eight-story Blockade Runner Hotel in Wrightsville Beach produced an unofficial reading of 138 mph. A more reliable measurement of a

112-mph gust was recorded at Kure Beach at 2:20 A.M. on September 16, just moments before the eye of the storm passed over. The automated Frying Pan Shoals station, located about thirty miles southeast of Cape Fear, reported sustained winds of near 100 mph for a twenty-minute period.

Other reports of peak gusts included 96 mph at the Corps of Engineers research pier at Duck, 91 mph at Cape Lookout, 86 mph at the Wilmington National Weather Service office, 82 mph at Cherry Point, 80 mph at Oak Island, and 71 mph at Myrtle Beach. In addition to these fixed weather stations, a special mobile research team was positioned on the coast to gather data and study Floyd's impact. The University of Oklahoma's Doppler-on-Wheels team set up its portable instruments on Topsail Beach, where team members recorded sustained winds of 96 mph and gusts to 123 mph during the early morning hours of the sixteenth. The lowest barometric pressures reported in North Carolina were 28.32 inches at New River Air Station near Jacksonville and 28.34 inches at the Wilmington Airport.

Floyd's lessened intensity at landfall also helped reduce the overall impact of tidal surge on the North Carolina coast. According to a report from the Corps of Engineers, Oak Island in Brunswick County was the hardest hit, suffering a measured surge of 10.4 feet. A measurement of 10.3 feet was taken on the sound side of Masonboro Island—10 inches less than the level recorded after Fran. Tides ran 9 to 10 feet above normal from Fort Fisher northward, and 6 to 8 feet above normal on the beaches of Onslow and Carteret Counties. Beach erosion was significant along the entire coast, with some barrier islands faring better than others. And as with most hurricanes, it wasn't just the ocean beaches that experienced high water. The Neuse and Pamlico Rivers and the Pamlico Sound were hit by a surge of 6 to 8 feet.

As the storm's most potent rainbands swept over eastern North Carolina, radar screens up and down the coast lit up with the bright greens, yellows, and reds that signify intense weather. In particular, forecasters monitored their instruments for the early signs of tornado development—and there were plenty of such signs to keep them busy. The Newport office of the National Weather Service issued a total of twenty tornado warnings, and ten twisters were verified by spotters in the area. Most of these came during the period from 4:00 P.M. to midnight, several hours before the eye of the storm reached land. One of the stronger tornadoes struck Hobucken in Pamlico County. The Weather Service rated it as an F2 on the Fujita Scale used by meteorologists to measure tornado severity. The Hobucken twister heavily damaged or destroyed several structures, including a mobile home, a house, three churches, and a school. Another tornado in Emerald Isle destroyed two houses and heavily damaged three others. Four tornadoes were spotted near Wilmington, and others were confirmed in Bertie and Perquimans Counties.

Without question, Floyd's winds, tides, and tornadoes were dramatic and

destructive, particularly along some portions of the coast. But in more than two dozen counties, residents experienced an even greater threat that would later prove the catalyst for Floyd's most disastrous legacy. This was, of course, the unending rain that accumulated in ditches, creeks, and rivers, eventually swelling into the great flood. The rains, which in some areas lasted more than sixty hours, fell over a region whose soils were already saturated beyond capacity by the earlier downpours during hurricane Dennis. River levels were already high when Floyd's first steady showers began to fall. As they gathered continuously updated reports of rain accumulations across the region, Weather Service officials and emergency planners began to issue a steady flow of around-the-clock flood watches and warnings.

The excessive rainfall that swamped eastern North Carolina before and during Floyd was the result of the storm's massive size and the timing of its approach. A strong trough of low pressure had moved in from the west as the hurricane came ashore, wringing extra moisture from the atmosphere and dumping it over land. Hurricane researcher William Gray described the trough's influence to the *News and Observer*: "The same kind of thing happened during Hurricane Hazel in 1954. It is a particularly bad weather pattern that can transform a relatively weak hurricane [into] a terrible one, and a terrible one into a catastrophic one."

At the peak of the Floyd flooding, Department of Transportation officials reported that 1,400 North Carolina roads were impassible. (Photo by John Althouse; courtesy of the Jacksonville Daily News)

Because rains began well in advance of Floyd's eventual landfall, total storm accumulations were very large. At Wilmington, rainfall was measured at 13.38 inches on September 15, establishing a new one-day record for that location. A new twenty-four-hour record was set as well when 15.06 inches fell between 3:00 A.M. on the fifteenth and the morning of the sixteenth. The storm total for Wilmington was a remarkable 19.06 inches. According to Weather Service reports, rainfall totals averaged between 14 and 16 inches across many inland counties, but some locations may have topped 23 inches. Other extreme measurements included 19.01 inches in Bladen County, 16.52 inches in Brunswick County, 16.06 inches in Myrtle Beach, 15.65 inches in Rocky Mount, 13.80 inches in Zebulon, 12.99 inches in Tarboro, and 12.86 inches in Greenville.

Thanks to the combined rains of Dennis and Floyd, several locations in North Carolina also established new one-month precipitation records. Wilmington registered 23.45 inches for September, beating out the old single-month record of 21.12 inches set way back in July 1886. September was also the wettest month on record in Raleigh-Durham, where the measured 21.80 inches broke a 112-year-old record. This total was well above the 16.65 inches that fell in September 1996 when hurricane Fran rolled through the Triangle. Other cities across the east also endured astronomical monthly rainfall totals. Snow Hill in Greene County received 35.29 inches in September, Greenville had 27.36 inches, and Kinston had 23.03 inches. Normal monthly averages for this region are around 5 inches. In addition to these hefty totals, Floyd surpassed other rainfall records in Pennsylvania, New York, and Connecticut.

The magnitude of the flooding disaster became apparent in the hours and days after the heaviest rains passed. Because there was no place for water to drain through already saturated soils or overfilled rivers, floodwaters began to back up into streets, homes, farms, businesses, and interstate highways. The waters rose quickly in some areas, more slowly in others. Many of the flood victims were taken by surprise—some were asleep in their beds when they were awakened by the sensation of water on their backs. Others endured the worst flooding days after the storm when rivers finally crested. Along Floyd's path and across at least a dozen Tar Heel counties, an epic flood held the entire population in its clutches.

Within days, media reports began to refer to the disaster as a "500-year flood." The waters were the highest in the memories of just about everyone who lived near the banks of eastern North Carolina's rivers and creeks. Not surprisingly, several reporting stations along the Neuse, Tar, and Northeast Cape Fear Rivers established all-time flood records in the days following Floyd. At Kinston the Neuse River crested at 27.71 feet on September 23—

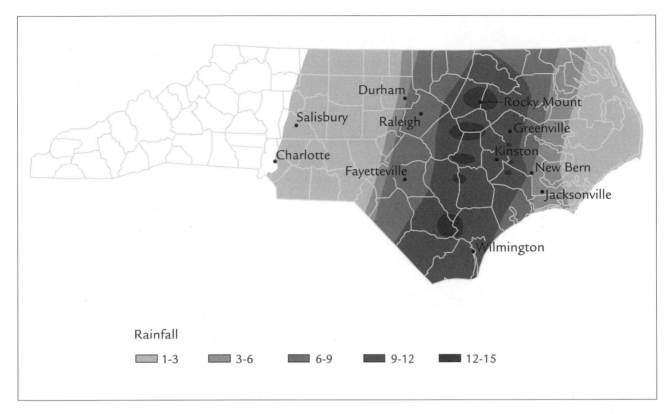

Rainfall

1-3	3-6	6-9	9-12	12-15

Hurricane Floyd storm-total rainfall. (From Joel Cline, "Surface-based Rain, Wind, and Pressure Fields in Tropical Cyclones over North Carolina since 1989," National Weather Service, Raleigh, N.C.)

more than 13 feet above its flood stage. This eclipsed the previous record of 25 feet set in July 1919. The new record mark was also more than 3 feet above the peak flood levels of hurricane Fran. Kinston and much of the Neuse River basin suffered through severe flooding for over three weeks. Water levels did not drop below 20 feet until October 10. The river was still above flood stage when hurricane Irene's rains fell on October 17, one month after Floyd's passage.

The Tar River crested in Rocky Mount at 32.35 feet, more than 17 feet above flood stage. On September 21, the Tar crested in Greenville at 29.72 feet, more than 16 feet above flood stage. The flooding in Greenville surpassed by more than 5 feet the previous record, set in 1919. At Chinquapin, the Northeast Cape Fear River also set a new record, rising to 23.51 feet on September 18. This was more than 7 feet above the river's crest during Fran. In all, at least nine U.S. Geological Survey river stations in North Carolina set new flood records, along with one in Virginia, six in New Jersey, one in Delaware, and two in

DATE	LOCATION	FLOOD STAGE (FT)	OLD RECORD	NEW RECORD
9/17/99	Tar River at Louisburg, N.C.	20	25.34	26.05
9/17/99	Tar River near Rocky Mount, N.C.	15	23.67	32.35
9/18/99	Fishing Creek near Enfield, N.C.	16	17.72	21.65
9/20/99	Tar River at Tarboro, N.C.	19	31.77	41.51
9/21/99	Tar River at Greenville, N.C.	13	22.07	29.72
9/17/99	Neuse River near Clayton, N.C.	9	20.12	20.67
9/20/99	Neuse River near Goldsboro, N.C.	14	26.21	28.85
9/23/99	Neuse River at Kinston, N.C.	14	23.26	27.71
9/18/99	Cape Fear River at Chinquapin, N.C.	13	20.16	23.51
9/17/99	Nottoway River near Sebrell, Va.	16	24.43	27.01
9/17/99	Saddle River at Lodi, N.J.	5	12.36	13.94
9/17/99	North Raritan River in Raritan, N.J.	10	15.51	18.78
9/17/99	Raritan River in Manville, N.J.	14	23.80	27.50
9/17/99	Millstone River at Blackwells Mills, N.J.	9	18.68	20.97
9/17/99	Raritan River at Bound Brook, N.J.	28	37.47	42.13
9/17/99	Rahway River at Springfield, N.J.	5.5	9.76	10.67
9/17/99	Christina at Coochs Bridge, Del.	9	13.12	13.92
9/17/99	E. Brandywine near Downingtown, Pa.	7	13.40	14.74
9/17/99	Brandywine Creek at Chadds Ford, Pa.	9	16.56	17.15

Source: National Weather Service/U.S. Geological Survey.

Pennsylvania. And of the more than two hundred stream gages monitored in North Carolina, twenty were destroyed during the storm.

Initial data from the USGS indicated that at least eleven of its monitoring sites in North Carolina exceeded 500-year flood levels. This term is used by hydrologists who use statistics to anticipate flood recurrence intervals. For a particular location, a 10-year flood has a 10 percent chance of occurring in a given year. A 100-year event has a 1 percent recurrence rate, and a 500-year flood has only a 0.2 chance of happening in a given year. But preliminary USGS data suggested that 500-year flood levels were reached in Ahoskie, Rocky Mount, Hilliardston, White Oak, Enfield, Tarboro, Lucama, Hookerton, Trenton, Chinquapin, and Freeland. Months after the storm, however, those 500-

year flood estimates were revised by officials at the USGS. After reviewing many outdated and inaccurate flood maps, they determined that Floyd may have been more of a 150- or 200-year flood.

But regardless of the flood's eventual place in history, its severe effects are indisputable. The high water covered so many roads, bridges, and highways that entire communities were cut off from the outside world. By the time the rivers crested, nearly all major roads in Duplin, Jones, Pender, Greene, Lenoir, Craven, Pitt, and Edgecombe Counties were impassible. Among the major transportation arteries blocked by flooding were U.S. 70 in Kinston, U.S. 17 in Washington, Pollocksville, and portions of Brunswick County, U.S. 264 between Greenville and Washington, U.S. 64 in Edgecombe County, U.S. 24 in Beaulaville and Kenansville, I-40 in Pender County, and I-95 near Rocky Mount. At the peak of the flood, Department of Transportation officials reported that 1,400 eastern North Carolina roads were impassible. Bridges washed out in dozens of places and asphalt crumbled in others, creating submerged driving hazards that were invisible to motorists. Across eastern North Carolina, 283 bridges were damaged by the storm, and 17 were completely washed away or undermined. The impassible roads complicated the disaster by slowing relief efforts in many cities, forcing drivers to make elaborate detours and delaying the delivery of food and supplies for days. Five shelters — three in Edgecombe County and two in Pitt County — were completely isolated by flooding and could only be supplied by helicopter. By late October, weeks after the storm, 232 of the state's roads were still closed to traffic.

Tragically, the submerged roads proved to be the most dangerous places to be in the disaster — they were responsible for most of the storm-related deaths in North Carolina. People tended to underestimate the depth of the water, the swiftness of the currents, and the risks associated with trying to drive over roads they couldn't see. Some of these dynamics were described in a report in the *Virginian-Pilot*, with help from FEMA and the American Red Cross:

> Water weighs 62.4 pounds per cubic foot and typically flows downstream at 6 to 12 miles per hour. When a vehicle stalls in the water, the water's momentum is transferred to the vehicle. For each foot the water rises, 500 pounds of lateral force are applied to the vehicle.
>
> The biggest factor, however, is buoyancy. For each foot the water rises up the side of the vehicle, it displaces 1,500 pounds of water. In effect, the car weighs 1,500 pounds *less* for each foot of rising water around it. In other words, 2 feet of water will carry away most vehicles.

Submerged roadways were only part of the problem. Within hours of the storm's arrival and the subsequent flooding, thousands of residents in more than a dozen Tar Heel counties found themselves trapped in their homes by rising water. Many were shocked to find the coffee-colored fluid edging into their yards—especially those who had never seen floodwater in their neighborhoods at all. Along with the rising waters came a witches' brew of leaked gasoline, sewage, and other contaminants that spread a pervasive stench through virtually every flooded area. Thousands were forced to flee their homes and businesses and escape to nearby high ground—if they could find any. Many of those ousted by the flood made their way to rooftops and treetops to await rescue. Across eastern North Carolina, thousands of men, women, and children, along with dogs, cats, and other family pets, waited for hours on steeply pitched roofs that became islands in a sea of muddy water. In countless neighborhoods, floodwaters buried cars and trucks and most anything else less than eight feet tall. Homes in Greenville, Tarboro, Princeville, and Rocky Mount were submerged beyond recognition, with waters in some locations standing more than fifteen feet deep.

Rescue teams formed quickly and fanned out across more than a dozen eastern counties. The task of rescue coordination was massive and involved state emergency managers, the National Guard, the U.S. Coast Guard, the U.S. Navy, and virtually every division of the State Highway Patrol, county sheriff departments, local police, and fire and emergency service personnel that was available. National Guard units from as far away as Georgia, Kentucky, and Texas arrived to provide aid. Hundreds of volunteers joined forces with government officials to scour the countryside in search of stranded flood victims. Through long days and nights they patrolled neighborhood streets and rural areas in bass boats and small skiffs, looking for isolated families and stranded individuals. Nearly fifty rescue helicopters were put into service by various branches of the military, focusing primarily on Rocky Mount, Princeville, and portions of Jones, Pitt, and Duplin Counties where flooding was worst. At one point, there were so many choppers in the skies over Rocky Mount that one Coast Guard helicopter hovered high above the rest, doing nothing but air traffic control.

At the Army National Guard Logistical Staging Area in Kinston, more than a thousand guardsmen helped carry out hundreds of flight missions in the first days after the storm. In addition to the rescue of stranded storm victims, they also helped deliver food, water, and medical supplies to shelters and neighborhoods cut off by flooding. In Greenville, choppers lifted hospital employees to a medical supply warehouse that was isolated by the flood. There the workers were lowered to the ground, where they retrieved nearly two dozen large boxes of doctors' bags and other supplies for later distribution to hospitals and clinics around the area.

The dramatic rescues during this time were unlike any others in North Carolina's hurricane history. By midafternoon on September 17, almost 1,500 stranded residents had been plucked from rooftops and treetops by rescue teams. Among them were about 500 people in Jones County, 500 in Duplin

A marine rescue helicopter airlifts a desperate trucker from the cab of his rig after floodwaters covered both lanes of Interstate 95 in Nash County. (Photo courtesy of the Raleigh News and Observer*)*

County, 300 in Edgecombe County, and dozens more in Beaufort, Pitt, Craven, and Greene Counties. At least 400 of those rescued had been lifted out by helicopter, some in the black of night. One man held on to the branches of a shrub for nine hours, screaming for help, until rescuers finally reached him. "It was the most terrifying experience I ever had in my life," Dunbar resident and evacuee Leatha Norman told the Associated Press. "I felt like I was going to lose my family and my life."

But as the hours rolled by, reports filtered in that thousands more were stranded and in need of rescue. By Saturday, September 18, officials in Edgecombe County reported that more than 3,500 residents had been rescued in

Lonnie Smith, a twenty-three-year Marine stationed in New Orleans, drove over 1,000 miles toward Hubert to check on his family before floodwaters stalled his minivan on N.C. 258. Two fellow Marines used their truck to pull Smith from the flood. Driving in floods is hazardous, and well over half the fatalities attributed to Floyd in North Carolina involved motorists attempting to cross submerged roads. (Photo by Don Bryan; courtesy of the Jacksonville Daily News)

that county alone. Renee Hoffman, spokeswoman for the state's emergency management office, described the process for the *News and Observer*: "It is neighbor rescuing neighbor in a lot of situations out there. There is no way we have a firm number of people who are waiting. We assign a flight to pick up a certain number of people and they find others along the way. This is a find-people-as-you-find-them situation." In many locations, pilots donned night-vision goggles and continued to fly round the clock in search of stranded victims. After nineteen straight hours of rescue missions, Coast Guard Lt. Robert Keith told the *Virginian-Pilot*: "We'd be flying to one spot where the sheriff said someone was waiting for us, and we'd fly over all of these people — rows and rows of people with flashlights, waiting on their rooftops, click, click, clicking for us to come help."

Some motorists on I-95 in Nash County were among those rescued in dramatic fashion. Sherry Boyer, a Pennsylvania resident who was traveling the interstate on the morning after the storm, died of a probable heart attack when her van stalled in a four-foot-deep torrent that covered the highway. A motorist attempting to assist her waded into the current with a heavy chain to attach to her van but was swept off the highway into deep water. Jim Howell, a local volunteer firefighter who happened on the scene, watched as the man was swept over the highway and into the woods. Using his handheld radio, Howell called for help, and within minutes a twin-rotor Marine helicopter was lowering a Marine to rescue the man. Soon after the man was pulled to safety, Boyer's van was recovered, but attempts to revive her were unsuccessful. Then, as Howell and others were still gathered on the edge of the flooded highway, several truckers attempted to cross the deepening stream that cov-

On dozens of roads across eastern North Carolina, runoff from Floyd's imposing rains washed away bridges, culverts, and large sections of asphalt. At this washout near Richlands, water lines, phone lines, and cable TV were also severed. (Photo by Don Bryan; courtesy of the Jacksonville Daily News)

ered both north- and southbound lanes. Within minutes, two large trucks lost control in the shoulder-deep water, drifted sideways into the median, and stalled. A second Marine chopper soon arrived and hoisted the truckers to safety, one by one.

Several of the helicopter crews involved in rescue missions were later honored by the Army and Navy for their heroism during the floods. They were recognized for saving scores of lives, for dropping rescue swimmers into dangerous areas, for flying as close as five feet from high-tension electrical lines, and for the great compassion they showed for the victims they rescued. Never before had missions of this kind been deployed on such a large scale in North Carolina. Though many of the rescue teams had acquired their expertise in places like Kuwait and Nicaragua, they were clearly moved by the destruction they witnessed in their own country. Many of the victims they rescued were also emotional about their airborne evacuations. One young Greenville resident even decided to join the Navy after his family was saved by a hovering chopper. Derek Latham, a senior at J. H. Rose High School, enlisted after watching his mother and two sisters pulled to safety from their flooded home. "That was the moment that I decided joining the Navy was what I wanted to

do," Latham told the *News and Observer*. "The only thing I would like to do is the same thing they did for my family."

Many of the storm's most daring rescues were also linked to tragedy. As rains poured over eastern North Carolina on September 17, Mitchell Piner and his fifteen-year-old stepson, Gary Williams, were on their way to visit a friend when their truck was swept off N.C. 41 just outside Wallace. The swollen waters of Rockfish Creek formed a strong current that smashed the truck into a nearby tree. Piner and Williams managed to crawl out of the cab through the back window, but both soon struggled to keep their heads above the swiftly moving waters. Erasmo Mencias, a fifty-one-year-old construction worker from nearby Magnolia, witnessed the truck's demise and jumped into the raging flood when he heard shouts for help. He managed to find Williams and grab him in a bear-hug, inching him up a tree to wait for more rescuers. While the creek waters continued to rise, Mencias held the teenager tightly for more than thirty minutes until the Wallace Volunteer Fire Department arrived with a boat. Mencias had saved Williams's life, but unfortunately Piner became one of Floyd's many drowning victims. For Mencias, a strong swimmer who used to set lobster traps off the coast of Honduras, saving this boy from drowning was not a first. He had rescued four people in a similar way during the floods that followed hurricane Fifi in 1974.

Another of the storm's dramatic rescues was reported by the *News and Observer*:

Pender County, along I-40, 4:30 P.M.: Burgaw Creek, Rockfish Creek, and other streams washed over one of Eastern North Carolina's main arteries.

Powerful currents of knee-high water streamed across the highway, flooding dozens of cars and sweeping at least one Pender County motorist away, apparently to his death.

Highway Patrol Commander R. W. Holden, the agency's top cop, waded along the highway Thursday afternoon near Wallace, ordering eastbound motorists to turn back. "It's too dangerous," he told them. Behind Holden, a rescue effort was under way. The driver of a minivan tried to cross the flooded westbound lanes near Wallace. Witnesses said the van was swept off the shoulder into a ditch that Rockfish Creek had turned into a torrent.

The roof of the van disappeared under the water. Another motorist, Matt Wilde of Wilmington, dove into the water to try to save the driver. But the waters quickly swallowed Wilde, too. For more than a half-hour, Wilde clung to a tree while troopers tethered to a truck swam in to get him.

Troopers tried several times to get a rope around Wilde and swam under the waves in an attempt to reach the van. Highway Patrol Sgt. Terry Carlyle tried to smash the window of the van with a crescent wrench, but the current was too strong and he couldn't break the van open. Other motorists ran down the highway to volunteer their arms and backs in the rescue effort.

At 5:20 P.M., the rescuers pulled Wilde out of the current. Carlyle hugged Wilde, who was shaken but uninjured. "You did a great job," Carlyle told him as Wilde gazed back at the water where the van had disappeared, its driver almost certainly drowned.

In dozens of communities, emergency workers and volunteers ventured door to door in small boats in search of marooned flood victims. They rounded up fleets of bass boats, aluminum skiffs, canoes, inflatable zodiacs, and even Jet-Skis to search submerged neighborhoods and rural homesteads. In places like Rocky Mount, Tarboro, and Princeville, where the flooding was most extreme, rescuers described steering small boats over the tops of submerged backyard fences, clotheslines, storage sheds, and swing sets. The flooding was so deep that some reported bumping their engine propellers on the tops of sunken pickup trucks. Another rescue worker recognized the blue lights atop a sheriff deputy's car "at least two feet" below the murky water. Those searching for stranded victims usually didn't have to travel far, and countless runs were staged day and night to deliver families and their pets to safety on higher ground. According to *MSNBC News*, rescuers in Duplin County fanned out on Jet-Skis, and their wakes "helped scare off deadly water moccasins that wriggled through flooded windows of mobile homes."

Emergency workers and volunteers tirelessly repeated their efforts for days after the storm. Most were very successful. One particular rescue mission on the night of the hurricane failed, however, becoming the single greatest tragedy of the Floyd disaster.

As the hurricane's drenching rains swamped Edgecombe County on the night of September 16, residents in the small community of Pinetops watched the waters rise into their streets and yards. Even in the dark of night, many could see that their homes would soon be filled with the overflowing waters of Town Creek. Ben Mayo, who loved to fish the creeks near his home, scrambled to launch his flat-bottomed boat to rescue his neighbors and family. After collecting eight other people and transporting them to safety, Mayo returned to his home amid the storm's pounding winds and rain. There he loaded his wife, daughter, granddaughter, and several neighbors into his small boat for the short ride to higher ground. But with twelve people onboard, the small skiff soon capsized in the shifting currents of the flood. Tossed into the pitch-black water, only half of the boat's passengers were able to scramble to the safety of a nearby bank. Drowned in the flood were Mayo, his wife Vivian, his daughter Keisha, his five-year-old granddaughter Teshika Vines, and two young neighbors, Cabrina and Destiny Flowers. Mayo, who was well respected in the Pinetops community, had died while trying to help others survive the storm.

Though helicopters, boats, and National Guard transport trucks were the vehicles most often used to retrieve stranded flood victims, one Pitt County family found a resourceful way to save its neighbors. Charles Davenport of Pactolus took the advice of his son and used his Hagie crop sprayer to maneuver through high water and muck to rescue about sixty-five residents. The sprayer, which features a six-foot ground clearance, managed the high water without problems since the engine and power train were above the flood. Davenport's stepson, Chris Sawyer, backed the machine—its spray arms folded upward—up to his neighbors' front porches and offered rides to high ground. Driving the machine was sometimes treacherous, but Sawyer moved slowly and cautiously. Eventually, flood currents became too swift for safe operation of the vehicle, but by then most of the neighbors had been transported to safety.

One of the most dramatic rescues of hurricane Floyd took place a full day before landfall, in the open Atlantic some three hundred miles east of Jacksonville, Florida. The eight men aboard the oceangoing tug *Gulf Majesty* knew that the storm was on their heels but made the mistake of thinking they could outrun it. On the morning of September 15, the tug's crew cut loose its 750-foot barge when waves the size of four-story buildings flooded the tug's engine room. After notifying the Coast Guard by radio, the crew abandoned the sinking ship and launched into heavy swells and 60 mph winds. As the men clambered into their bright orange life raft, a wave dashed the raft away from the tug, separating three of them from the five who had made it safely aboard. With their tug now rapidly foundering, the three had no choice but to jump into the raging Atlantic.

Meanwhile, about 140 miles away, huge swells broke over the flight deck of the aircraft carrier USS *John F. Kennedy*. Along with a small fleet of other Navy ships, the *Kennedy* was at the time moving out to sea to avoid the wrath of hurricane Floyd and its near-category-five winds. Having a clear fix on the *Gulf Majesty's* emergency locator beacon but no ships in the area, the Coast Guard radioed the *Kennedy* to ask for help in rescuing the tug's hapless crew. The massive carrier turned back toward the approaching storm and readied a pair of SH-60 Seahawk helicopters for the mission. The two choppers sped toward the lost ship and zeroed in on the emergency beacon. When they arrived, they found the three men in the water desperately clutching a broken broomstick to stay together. Then, as if taking his cue from the pages of Sebastian Junger's *The Perfect Storm*, Navy rescue swimmer Shad Hernandez jumped into the raging seas amid thirty-five-foot swells to save the three men. The rescue went smoothly, and within eleven minutes, Hernandez and the three crewmen were safely on their way back to the *Kennedy*.

After returning to the ship to refuel, the helicopters turned back once again to search for the five crew members still at sea. The men were later found clinging to their raft, eight hours after their tug had gone down. Storm-beaten and exhausted, they had barely held on to the raft as it rode up and down mountainous waves that sometimes folded it in half. Once again, Navy rescue swimmers went into the water and successfully hoisted the men to safety. Of the many helicopter missions that took place during or after hurricane Floyd, few were as daring as the rescue of the *Gulf Majesty's* crew.

The storm's unprecedented flooding across a broad portion of eastern North Carolina brought misery and destruction to dozens of cities and towns. Among the hardest hit was the Edgecombe County town of Princeville, where floodwaters were perhaps deeper and the devastation more complete than in most other locations. Princeville, which had been founded on the banks of the Tar River by former slaves after the Civil War, was submerged when floodwaters burst through a protective dike in seven places. The town of 2,100 residents was quickly swamped by waters that measured close to thirty feet deep in some locations. The inundation buckled roofs, stacked cars on top of one another, and swept at least a dozen houses off their foundations. Surviving homes and businesses were coated with a heavy layer of foul-smelling mud that later proved nearly impossible to remove. More than six hundred homes were heavily damaged.

One of the challenges faced by town residents was the recovery of hundreds of caskets unearthed by the flood. The sealed coffins floated out of their graves and were swept about the town, along with propane tanks, appliances, and other household flotsam. While the floodwaters were high, crews working from boats gathered the wayward caskets and tethered them to trees and poles. Two weeks after the storm, 129 caskets had been recovered and taken to a temporary morgue, where a team of forensic experts made the necessary identifications. The morticians were aided by relatives who supplied information about clothing items and jewelry worn by the deceased. Ultimately, some 224 caskets were recovered, of which 174 were positively identified. All were later reburied. Elaine Wathen of the N.C. Division of Emergency Management told the Associated Press: "It's something we've never dealt with before. We wanted to be sure we did everything right and had the utmost respect for the remains."

By the end of September, most of the water had receded from the tattered streets of Princeville. But few residents were able to return to their homes due to the extent of the destruction. And just when it seemed things couldn't get any worse, a wave of looting hit the town. Thieves made off with truckloads of stolen items from more than a dozen flood-damaged homes. Princeville's small police force, which had lost two of its three patrol cars to the flood, sought help from FEMA to pay for four temporary officers to provide additional security. About this time, disheartened residents were given a boost when civil rights leader Jesse Jackson and American Red Cross president Bernadine Healy visited the devastated town. Jackson delivered a rousing speech to the three-hundred-plus residents who were still sleeping in a local gymnasium. Though the waters were mostly gone, it would be many more months before residents would return to their homes. According to the *News and Observer*, Jackson lifted spirits by mixing "poems, prayers, and promises in his trademark style." Jackson chanted to the crowd: "We can make it. Say after me: 'I am somebody.' Through the rain, through the flood: I am somebody. We can make it. We can make it. Keep hope alive. Keep hope alive."

For several months after Floyd, Princeville served as a focal point for the many issues surrounding poststorm recovery and reconstruction. By November, town officials were still wrestling with the question of whether to rebuild the protective dike and replace damaged homes or to participate in federally sponsored buyouts. The buyouts offered by FEMA would have provided funding to relocate many of the town's residents to other neighborhoods out of the 100-year floodplain. If that option had been selected, it would essentially have broken up the historic town. After extensive debate, the Princeville Board of Commissioners voted 3–2 to ask the Corps of Engineers to repair its dike. Construction was completed by the summer of 2000, and many of the flooded homes were repaired.

On September 18 in nearby Tarboro, dozens of residents walked or rode bikes into town to watch the Tar River crest about twenty-two feet above its nineteen-foot flood stage. As in other towns built on the banks of eastern rivers, Tarboro's downtown business district was submerged to a level that far exceeded any in memory. And, like people in so many other communities hit by the flood, hundreds of Tarboro residents were rescued by boat and heli-

copter in the first days after the storm. Many who escaped said the waterline was "just above the light switches" in their homes. They gathered in emergency shelters overflowing with despondent families who could hardly believe what was happening. Initially, some two thousand refugees filled Tarboro High School, which had no running water and no electricity. As the hours and days crept by, food, drinking water, and personal effects were in short supply, and hot showers were nonexistent. But for many, the greatest need was to know the whereabouts of relatives and friends. In many instances, men were rescued by helicopter hours after their wives and children had been taken to safety, so family members spent long nights searching for one another by sending messages through police and shelter officials. No one was keeping lists of who had been rescued and what shelter they had been taken to.

Edgecombe County officials set up a makeshift operations center near a jail where refugees were to be dropped off after being rescued by helicopter. But with the power out, Coast Guard and other military pilots had trouble seeing the improvised landing field in the dark. To solve the problem, sheriff's deputies parked their patrol cars around the field and flashed their blue lights to guide the pilots to the ground. Officers were called away, however, when reports came in that the K-Mart near Tarboro High School was being looted. When they arrived at the scene, they found hundreds of local residents sleeping in their cars in the oversized parking lot, which was on a patch of high ground. Apparently, some had broken into the store—not to steal merchandise, but to use the bathroom.

Shelters throughout the area were brimming with storm survivors. By midnight on September 17, more than five thousand people were sprawled across the gymnasium and cafeteria floors of several Edgecombe County schools. With access to television, radio, or any other news media very limited, accurate reports about the disaster were few and hard to come by. Along with their many other duties, shelter officials spent considerable time trying to quell rumors. As the days passed, word spread that a tuberculosis outbreak was under way and that several area shelters were set to close. Another rumor held that refrigerated trucks being guarded by the National Guard near a local hospital contained the bodies of flood victims. The trucks merely contained food.

On September 20, FEMA opened disaster recovery centers in a handful of the hardest-hit cities, including Tarboro. Lines formed quickly as flood victims waited to sign up for whatever aid was available. Those who still had telephone service registered via FEMA's toll-free phone line, while others could only register in person. As they stood in queues that stretched out into the street, they swapped painful stories about the flood and shared tips on where to find ice and fuel. Patricia Foreman, a home day-care operator whose East Tarboro home was destroyed in the flood, managed to keep her loss in perspective. After describing her own ordeal, she told the *News and Observer*: "But I know a woman who's really bad off. Her house is gone, her momma's house is gone, her sister's house is gone, and her babysitter's house is, too."

Among the many public buildings swamped by floodwaters was the Edgecombe County Courthouse. Water filled the courthouse basement and saturated 130 years' worth of legal records. After the waters were gone, county workers were left with the formidable task of recovering and restoring the damp and moldy papers that filled rows of file drawers. To accomplish the task, they sought help from state officials and a recovery contractor from Texas. After thousands of documents had been sorted through in a downtown Raleigh parking lot, the files were loaded into a refrigerated tractor-trailer and trucked to the contractor's facility in Fort Worth. There, they were freeze-dried and disinfected. Held under pressure at −20°F, the moisture in paper turns to ice, which then evaporates. After undergoing this process, the records were left in better shape than they would have been after normal air drying. The same procedure was used in other cities where books and important records were soaked by the hurricane's floods.

In Rocky Mount, the worst flooding in history spread throughout the town, overtaking major highways and businesses and inundating entire neighborhoods. Like so many of Floyd's victims, most of those who watched the waters creep into their homes had no flood insurance — they had been told they didn't need it. Warehouses filled with goods were buried in floods that covered their roofs. Stores and stockrooms were flooded with waters that soaked merchandise on the highest shelves. The Food Lion on U.S. 301 was submerged by five-foot-deep waters, and store managers assessed the damage from a small raft they maneuvered inside the store. Along with the residents of many other hard-hit cities in the east, the people of Rocky Mount scrambled to escape the rising water, climbing onto roofs and launching small skiffs in the darkness of the storm. Their memories of Floyd are filled with images of unrecognizable neighborhoods and the constant thumping sounds of choppers in the sky.

After the storm passed by and the cresting Tar River subsided, state officials moved quickly to create temporary housing for thousands of displaced flood victims. An undeveloped ninety-acre tract in an industrial park north of Rocky Mount was selected as the future home of hundreds of Edgecombe County refugees. In the days following the storm, workers labored around the clock to build gravel roads, install utilities, and prepare the site for more than three hundred travel-trailers. The "FEMA city" was operational by September 22, and exhausted storm victims were soon back in livable quarters. One of the first to move in was Princeville resident Mattie Jones, who had been staying in a school shelter in Tarboro since the storm. She told the *News and Observer*: "It may not look like much to other people, but to me it's a castle. First thing when I got the keys in my hand, I said 'Oh, thank you God, a house, a bathroom, a bed, quiet.'" At the time Jones moved into her trailer, her Princeville home was still underwater.

Downriver at Greenville, more record-breaking floods forced thousands out of their homes. Among those ousted were almost five thousand East Carolina University students living in apartments near the Tar River. Though flooding struck some parts of the city and left others untouched, it caused problems everywhere. Late on Friday, September 17, rising water near the river short-circuited a critical transmission station, cutting power to Greenville's 48,000 electric customers. On the following Tuesday, when the flooding reached its highest level, the city's water plant shut down and water was unavailable. Thousands initially sought refuge in Red Cross shelters across Pitt County, but eventually many were moved to temporary housing around the city. Some moved in with area residents and students whose homes were not flooded. National Guard troops arrived to assist with rescue efforts and took up residence on the East Carolina campus. The university suspended classes, parents' weekend was canceled, and the Pirate's nationally televised football game with the University of Miami was moved to Carter-Finley Stadium in Raleigh. Classes did not resume until September 29, almost two weeks after the storm.

Flooding in Greenville was far worse than anyone could remember. Some apartment buildings were flooded up to the second-floor level, and many structures were filled with murky water well above door-knob level. All around town, emergency workers in small boats rescued hundreds of isolated residents. More than six thousand homes suffered some flooding damage. When the waters finally receded, piles of furniture, carpets, and appliances filled side streets near the river. City engineers estimated that 50,000 cubic yards of trees and storm debris would need to be hauled away. Pitt County residents worked for weeks to clean up the mess. As in so many other communities across the region, Greenville residents spent long hours washing mud and mold from their walls, floors, and furnishings.

About the time the flooding was at its peak and utilities were shut down, an unexpected explosion rocked Greenville. On September 22, a gasoline truck outside Pitt Memorial Hospital went up in flames while refueling fire department pumper trucks. No one was injured in the blast, and the hospital did not catch fire. The fire department pumpers were at the time being used to run a makeshift water system for the hospital. Water from a rehabilitation pool was used to keep toilets flushing and other equipment functioning, while bottled water was stockpiled for drinking. Like thousands of other area residents, hospital staff made the best of a very difficult situation.

As the flooding worsened across the eastern counties, some residents became desperate. After a National Guard convoy delivered drinking water to Greenville on Sunday, September 19, a second convoy was scheduled to deliver food and water on the following day. The unarmed crew ran into trouble, however, when the convoy was chased down by carloads of anxious residents who pulled out guns and baseball bats, demanding food and water. The police were later called in, but provisions were left with the residents, some of whom had not eaten for days.

Communities all along the Neuse River were under siege because of flooding. Waters rose to record levels in portions of Goldsboro, and flooding in Seven Springs swamped all but four of the town's ninety-four homes and businesses. In Kinston, the flooding far surpassed that of hurricane Fran, bringing diesel-fouled floodwaters into hundreds of homes, shops, motels, and restaurants. Rising more than thirteen feet above flood stage, the Neuse spread beyond its banks and crept over U.S. 70, forcing that major artery to close between its intersections with N.C. 55 and U.S. 258. Two bridges were undermined, and many in the city were isolated from the outside world. Though some towns and cities experienced flooding that lasted perhaps a week, Kinston's high waters endured for more than three weeks.

When the overflowing Neuse first surged across U.S. 70 on September 21, two nearby hotels were filled with customers—many of whom were evacuees from other areas. Before the rising waters inched above the hoods of their cars, many of the motel guests at the Days Inn and the Super 8 fled for higher ground. Later, three National Guard trucks plowed through the deepening waters of the Super 8 parking lot to urge a few remaining residents on the second floor to leave. Some refused to go, while others were carried to local shelters. Other businesses along the U.S. 70 corridor were similarly swamped. Landmarks like the Neuse Sport Shop and King's Barbeque were filled with muddy floodwaters that reached well above tabletop level.

Though cities and towns along the Tar, Neuse, and Northeast Cape Fear Rivers suffered the most during Floyd, dozens of other communities across eastern North Carolina were affected as well. In all, the National Weather Service reported that the downtown areas of at least thirty North Carolina towns were flooded. The heavy damages were not limited to communities along major rivers but included towns like Vanceboro, flooded by Little Swift Creek; Grifton, flooded by Contentnea Creek; Windsor, flooded by the Cashie River; and Trenton and Pollocksville, flooded by the Trent River. Across Jones County, where flooding in Trenton was "the worst in memory," almost a thousand people were housed in five shelters that included schools, a church, and a former topless bar.

In Windsor, the normally placid Cashie rose about a foot every two hours until it finally swamped virtually every business along King Street, the town's main drag. According to the *Virginian-Pilot*, at Jake's Barber Shop, where Jake Mitchell has cut hair since 1943, the flood "climbed more than 8 feet up the walls, knocked out the front window and sucked two big wooden cabinets into the street." It trashed an IGA supermarket, flooded the town hall and county newspaper office, and filled two insurance offices with four feet of water. Local residents commented afterward about the surprising nature of the flood, noting that even the insurance agencies didn't have flood insurance.

But river flooding was not the only destructive force associated with Floyd. Along portions of the coast, tidal surges caused heavy damages, especially on

Oak Island in Brunswick County. Not since the awesome storm surge of hurricane Hazel in 1954 had this resort community suffered such destruction. The recently incorporated town, which includes Yaupon Beach and Long Beach, took the brunt of the storm as landfall was just a few miles to the east. The heaviest damages occurred along the ocean beach, where Floyd's ten-foot storm surge and battering waves knocked down scores of cottages, scuttled two fishing piers, and filled streets with deep sand. According to town officials, 40 houses were destroyed, 290 more were heavily damaged, and another 250 were in need of minor repairs. Throughout much of the community, homes built under the protection of oak and pine trees showed little evidence of Floyd's effects.

Along the hardest-hit stretch of the beach, several homes collapsed into the surf while others were completely missing. Half-buried debris was scattered about the strand, having been tossed out of demolished vacation homes. Mattresses, televisions, bicycles, and washing machines were among the recognizable objects that lined what used to be Beach Drive. The kitchen of the popular Windjammer Restaurant was knocked into the surf, and the town's oceanfront boardwalk and gazebo complex were gone. The Long Beach Pier, which used to claim the title of the state's longest fishing pier, was cut down some three hundred feet by crashing waves. The Ocean Crest Pier lost about five hundred feet.

On nearby Holden Beach, about five houses were destroyed and another fishing pier was heavily damaged. At the Brunswick Community Hospital in Supply, a large section of roofing was torn off during the storm, just when a pregnant woman was about to deliver. Rain water poured into the building's air-conditioning system, and the expectant mother had to be transferred out of the hospital amid high winds. This was the second time within a year that the hospital had lost its roof: hurricane Bonnie's high winds had peeled back the covering in 1998.

In Shallotte, flooding from the Shallotte River filled downtown streets to a depth of eight feet. At least three families were rescued from their rooftops by boat soon after the storm passed through. The flooding tore away a sixty-foot portion of N.C. 130 just south of town, and high winds ripped the steeple off the nearby First Baptist Church. Some Shallotte residents sat quietly on their porches and watched floodwaters fill their yards, rising only to fire shotgun blasts at water moccasins that occasionally swam by.

At Carolina and Wrightsville Beaches, damages were moderate — not nearly as extensive as those left in the wake of hurricane Fran. At the northern end of Wrightsville, erosion brought Mason Inlet to within about ten feet of the foundation of the imperiled Shell Island Resort, which had been temporarily protected by a wall of sandbags. Winds peeled away large patches of stucco on many of the island's homes and businesses, and soundside docks and piers

An obstacle course of sand and debris filled Beach Road on Oak Island on the morning after hurricane Floyd made landfall in September 1999. (Photo by Jim Harper; courtesy of the State Port Pilot)

were twisted and buckled by hefty tides. Some ground-level apartments were flooded by two feet of water.

In the days following Floyd, isolated damages were reported from dozens of other coastal communities. On the Cape Fear River below Wilmington, several barges loaded with munitions broke free from their docks at the Sunny Point Military Ocean Terminal and were scattered by the storm. One ran aground across the river at Fort Fisher and had to be unloaded and dislodged from the marsh. At Topsail Island, where recent hurricanes had caused extensive ocean overwash, Floyd again cut three new inlets across the fragile beach. Most of the manmade protective dunes on the northern end of the island were leveled by high tides. In some locations, dunes had already been swept away by hurricane Dennis, exposing homes and roadways to Floyd's thrashing waves. Onslow County experienced losses not only at North Topsail, but throughout other communities where swollen creeks flooded homes and businesses. In all, the county reported 118 structures destroyed, 202 with major damage, and another 471 with minor damage.

Up the coast in Carteret County, most of the communities that suffered extensive flooding during Dennis were spared a repeat disaster. The western beaches, though, were hit hard by an eight-foot storm surge that left cottages teetering on the edge of severely eroded dunes. In addition to being hit by a twister that struck Emerald Isle and damaged several homes, the island was battered by waves that destroyed piers, undermined septic tanks, and carved away several more feet of precious sand. Officials later reported that, across the county, 17 homes were destroyed, 275 suffered major damage, and another 610 had minor damage.

Perhaps as significant as the number of lost structures was the even larger number of oceanfront properties that were left dangerously exposed to future storms after Floyd. The hurricane's impact on the frontal dunes of Pine Knoll Shores, Indian Beach, and Emerald Isle was so dramatic that hundreds of homes and condos were left standing literally at the edge of the sea at high tide. With almost no protective dunes remaining in some locations, home-owners feared the loss of their investments should another hurricane or severe winter storm strike the area. Tourist officials worried that the island's altered beach, which was almost nonexistent in some areas at high tide, would discourage vacationers from choosing Carteret County as their destination. Beach nourishment projects, though they had successfully been used to protect most of nearby Atlantic Beach, soon became a topic of heated debate across the county. In November 1999, a bond referendum proposing county-sponsored funding of a nourishment plan for the beaches from Pine Knoll Shores to Emerald Isle was put to a vote. But, plagued by questions of economic fairness and geological viability, the bond ultimately failed, leaving local officials in search of alternative funding schemes.

Hurricane Floyd's catastrophic effects were not limited to the Tar Heel state. As the storm rolled into eastern Virginia on the night of September 16, record rains fell across the Tidewater region, resulting in severe flooding, rooftop rescues, and property destruction not unlike that experienced in North Carolina. Some vehicles, swept from flooded highways, capsized and sank. Even one Virginia state trooper in Southhampton County had to abandon her cruiser and swim to safety when she was washed off of U.S. 58. At one point, more than three hundred roads across the state were closed. A fifty-foot section of road collapsed into Lake Powell when a dam near Williamsburg broke. Flooding was worst in Franklin, where nine to twelve feet of water filled the downtown district. Statewide, almost six thousand homes were damaged to some degree by the storm. In addition to at least one fatality linked to the flooding, two Virginians were killed by falling trees.

Though several mid-Atlantic states suffered through Floyd's blustery winds and heavy rains, massive rainfall over north and central New Jersey created another pocket of disaster far distant from the storm's original point of landfall. More than a foot of rain, accompanied by wind gusts over 60 mph, affected the region and extended into the suburbs north and west of New York City. But by far the area that experienced the greatest problems was Bound Brook, New Jersey, where the swollen Raritan River filled the business district with up to fifteen feet of water, forcing hundreds onto rooftops and setting the stage for still more dramatic rescues. The Raritan reached a record crest of forty-two feet, some twenty feet above flood stage. Working from small boats and helicopters, rescue teams pulled people from second-floor windows and rooftops throughout the day on Friday, September 17, and continued the treacherous work into the night. More than a thousand people were evacuated. Fires burned out of control in the business district, as firefighters could only watch the flames from aboard their Jet-Skis. Eventually, helicopters from New York arrived to drop water on the fires. A nearby water company was shut down, leaving tens of thousands of New Jersey residents without water for days. Across the state, 251 structures were destroyed in the storm and another 8,300 were damaged. Four residents were drowned in what officials described as "the single largest disaster to ever affect the state of New Jersey."

Back in North Carolina, emergency management officials struggled to deal with myriad problems, including a burgeoning health concern: what to do with the thousands, if not millions, of dead pigs, turkeys, and chickens drowned by the flood. In several eastern counties where large swine and poultry operations make up much of the agricultural economy, Floyd left behind a putrid mess. Carcasses of drowned hogs floated through submerged forests and farmlands, in some cases intermixed with a few exhausted animals that had managed to survive and gather on rooftops. From the air, windrows of pinkish hogs were visible below, and the numbers were astounding—early estimates suggested that upward of 30,000 swine had drowned. In addition, 2.4 million chickens and 700,000 turkeys died in the storm. This massive agricultural loss also posed a dangerous environmental threat that required quick attention.

State agricultural officials, working together with the Division of Emergency Management, arranged for large portable incinerators to be delivered to several eastern North Carolina farms where the animal problems were the worst. The incinerators, wood-fired metal boxes that resemble oversized garbage dumpsters, are capable of burning up to four thousand pounds of carcasses an hour. Crews gathered the dead animals from around the countryside, sometimes towing them behind small skiffs and piling them into large mounds with front-end loaders. Veterinarians and public health officials were present to monitor each burning operation. Over a period of several days, thousands of animals were collected and destroyed to minimize the risk to public health.

Though most of the dead hogs, turkeys, and chickens were gathered and burned in the first days after the flood, other serious environmental concerns lingered for weeks. In addition to flooding homes, cars, and businesses, the storm's rising waters had also submerged at least fifty hog-waste lagoons and twenty-four municipal waste-treatment plants covering a dozen counties. Chemicals of all kinds were flushed into rivers. Solvents, oils, paints, and drain cleaners were swept from hundreds of sunken homes. Millions of gal-

lons of sewage from the flooded hog lagoons flowed into the mix, causing some flood victims to become nauseous while they sat for hours awaiting rescue. At least five hog lagoons collapsed, including one in Duplin County that spilled an estimated two million gallons of waste into the Northeast Cape Fear River.

Flooded municipal waste plants caused similar problems by spilling untreated sewage into creeks and rivers in Cary, Smithfield, Hillsborough, Wake Forest, Zebulon, Kenly, Sanford, Maysville, Pink Hill, Fremont, Grifton, Goldsboro, Greenville, Wilson, Kinston, Jacksonville, and many other towns. In Wayne County, a ruptured dam carried with it at least two hundred feet of sewer line. Unknown quantities of waste from thousands of home septic tanks also leached into the flood. A variety of contaminants leaked from submerged automobiles, including gasoline, transmission fluid, and motor oil. Vehicles trapped in the flood were not the only culprits: dozens of flooded junkyards filled with rusting wrecks also contributed to the pollution. Industrial chemical spills were reported at six plants, including sites in Castle Hayne and Riegelwood.

The mixture of sewage, petroleum, and unknown toxins spread downstream with the floods and threatened the health of thousands. Because of the broad scope of the flood, the state's overfilled rivers carried with them a toxic soup of contaminants that spread wherever the waters went. Emergency workers began smearing Vick's Vapo-Rub under their noses to cope with the stench. From the air, chopper pilots commented on how the Neuse and Tar Rivers seemed to glow with an iridescent sheen that reflected light in a rainbow of colors.

Public health officials described the situation as "awful" and prepared for the worst. Those at greatest risk for gastrointestinal illness were the sick, the elderly, and the very young. Special precautions were taken to treat any such patients who had significant contact with floodwaters. Skin rashes were also a problem. In the weeks following the storm, stress took its toll on those most affected by the flood. Doctors saw a stream of patients whose illnesses were exacerbated by the disaster. Storm victims with asthma, high blood pressure, diabetes, and depression began seeking medical help in large numbers.

Water plant and well contamination was another area of grave concern, and residents in most affected counties were urged to boil their drinking water or to drink only bottled water. In the weeks following the storm, health officials began testing wells and municipal drinking-water plants as they came back online. By early October, the state's environmental health division was reporting that about one-quarter of the wells tested in the southeastern counties were tainted with floodwater and about 3 percent contained dangerous bacteria. But these tests did not include many of the wells in the most flood-ravaged areas.

(opposite)
Struggling to stay alive, hogs from a farm near Trenton wait for rescue on the roof of a swine barn as floodwaters from the Neuse River surround them. The following day, state officials called for reinforcements as they battled to save farm animals across eastern North Carolina from the waters of hurricane Floyd, the worst agricultural disaster in the state's history. (Photo by Mel Nathanson; courtesy of the Raleigh News and Observer)

Children in Snead's Ferry make the most of their Floyd experience. (Photo by John Althouse; courtesy of the Jacksonville Daily News*)*

Several of at least seven cars that were washed into the trees below the Walnut Creek dam near Goldsboro. (Photo by Kaye Farmer; courtesy of the Goldsboro News-Argus*)*

Even the Army National Guard fell victim to Floyd's treacherous floods. This transport truck had to be abandoned after it slipped into a ditch on Old Grantham Road in Goldsboro. (Photo by Sandra Renner; courtesy of the Goldsboro News-Argus*)*

Fortunately, the most immediate environmental concerns did not material-ize. No outbreaks of disease from tainted water were reported. According to area researchers, the rivers and sounds within the flooded area rebounded rapidly. Low oxygen levels killed only around 150 fish in coastal rivers. The state's fishing industry, which was expected to suffer terribly from the storm, actually fared well. The "dead zone" of low-oxygen, low-salinity water that had been present in the Pamlico Sound shortly after the hurricane dissipated within several months. Though state officials had predicted losses totaling millions of dollars in several fisheries, some fishermen instead reported near-record catches, and the commercial fishing industry had a good year overall. State officials tested fish and shellfish from several coastal areas and found no unsafe toxin levels. The entire industry fought false perceptions and worked to get the message out: North Carolina seafood was safe to eat.

After the floodwaters receded and the obvious pollutants had been cleaned up, environmentalists and state officials still questioned the long-term envi-ronmental impact of the storm. Groundwater, they felt, could be affected for years to come, and new testing programs were established. Water-quality mon-itoring was expanded across the eastern counties, including a program that uses state ferries to monitor algal growth and oxygen content in the Pamlico Sound. In December 1999, state legislators funded $27 million for environ-mental cleanup projects across the flooded region. Sites included 11 junkyards, 35 hazardous-waste sites, 277 old landfills, and 139 underground storage tanks.

Lawmakers realized very quickly that the Floyd disaster had amounted to an economic catastrophe for the state. After the storm, assessment teams fanned out to survey damage across the east. But with roads blocked and some neighborhoods flooded for weeks, totaling the damages was a time-consuming and difficult task. It was easy to see, though, especially from the air, that re-building eastern North Carolina was going to require massive amounts of state and federal aid. Of the thousands of flooded homes, state officials knew that very few were covered by flood insurance. President Clinton declared sixty-six North Carolina counties disaster areas, paving the way for federal re-lief. In the weeks following the storm, Governor Jim Hunt established the Hurricane Floyd Redevelopment Center to coordinate reconstruction efforts and to monitor the flow of money to devastated communities. He then pre-pared his request to the U.S. Congress: the state was seeking a whopping $5.3 billion in federal aid to get people back in their homes and fund the long recovery process.

But Hunt's proposed federal aid package was trimmed back and divided into two parts—a $1.7 billion request for immediate needs and $2.1 billion in long-term relief. Eventually, Congress came through with a total of around $2.2 billion, far less than had been expected. While lobbying for additional federal aid continued in Washington, immediate needs were not being met across the flood zone. This left Hunt and local lawmakers with little choice but to find additional money within the state. On December 15, 1999, the General Assembly convened a special session to address the funding shortfall. Out of the session came a special $836 million appropriation of state funds to provide safe housing, help farmers and small businesses, protect public health, clean up environmental hazards, and aid public schools and local governments. Of the $836 million, $504 million came "out of the hide of government" through budget cuts and canceled construction projects. Among the items cut was $6.7 million in planned renovations to the state's legislative buildings.

At the 2000 Water Resources Research Institute's annual conference, the governor's point man for the recovery, Floyd Redevelopment Center director Billy Ray Hall, gave the keynote address, in which he talked about Floyd's overall economic impact. Hall estimated that the disaster caused $5.5 billion in damages to North Carolina alone. A total of 63,000 homes were flooded, 7,300 were destroyed, and a quarter of a million people were displaced. Most significant, though, was Hall's assertion that it would take eight to ten years for eastern North Carolina to fully recover from the storm.

Officially, the National Hurricane Center places the cost of Floyd at $4.5 billion, or twice the amount of insured property damages. Still, many refer to Floyd as a $6 billion hurricane, and that figure rises even higher if lost wages, lost retail sales, and other tangent economic factors are considered. Early studies by the Economic Development Administration determined that the state suffered an economic loss of $4 billion due to the storm. According to the N.C. Division of Emergency Management, Floyd generated more than $3.6 billion in insurance losses and state and federal assistance. But the division's director, Eric Tolbert, added: "Systematically, we can't capture those uninsured or underinsured losses, which could easily match the known damages. When you include the known and the unknown, Hurricane Floyd will more than likely surpass Hurricane Fran in dollar damages. However, it is clear that Floyd had the state's most severe impact on all elements of disaster: economic, social, and health."

Floyd was particularly hard on farmers, and some small farm operators were at risk of losing their way of life after the storm. Agricultural losses were huge, totaling between $800,000 and $1 billion. State agriculture officials estimated that half the year's crops of peanuts, cotton, soybeans, and sweet potatoes were lost, along with 40 percent of the tobacco harvest. But more than just crops and livestock were destroyed. After surveying the damage and talking to state officials, Senate President Pro Tem Marc Basnight told *WRAL News*: "This is worse than the Great Depression. At least when the Depression was over, those folks still had their land, the vegetable plots, maybe a hog tied up out back. Many of these people, when this is over, will have nothing."

Fortunately for thousands of needy storm victims across the east, a massive outpouring of private support began flowing into the hardest-hit communities in the weeks following the storm. Churches, schools, civic groups, and families throughout the state — and from other states — collected truckloads of food and supplies for flood victims. Collection sites popped up everywhere, and soon leagues of volunteers were boxing, driving, and distributing goods to makeshift distribution centers in cities like New Bern, Kinston, and Rocky Mount. Nonperishable foods, diapers, water, clothes, and cleaning supplies were among the more popular items donated. Soon after the storm made landfall, Governor Hunt established the North Carolina Hurricane Floyd Relief Fund as a conduit for cash contributions. A statewide telethon raised $2.2 million, special collections made the rounds at Sunday church services, and individuals dug into their pockets to support the cause. In all, the fund received more than 66,000 contributions from around the world and raised over $19 million. This figure dwarfs the $70,000 raised following hurricane Fran in 1996.

Among the contributions were some large corporate gifts from a string of familiar names, including Food Lion, Blue Cross and Blue Shield of North Carolina, Lowe's Home Improvement Warehouse, R. J. Reynolds Tobacco, Glaxo Wellcome, Barnes & Noble, and Mount Olive Pickle Company. Dozens of other businesses donated cash, and many offered help with products and services. Stockroom Shoes gathered more than seven thousand pairs of shoes, National Gypsum Company donated $250,000 worth of wallboard, and grocery chains like Harris-Teeter and Food Lion gave away thousands of bags of ice and countless gallons of water. "The donations have been wonderful," distribution center volunteer Max Garrison said after spending several long hours unloading trucks and organizing goods. "They will help a lot. This food and these supplies won't just be nice to have, they will help these people survive."

After the floodwaters receded and the scope of the disaster could be better understood, the long, grueling process of rebuilding began. Throughout the fall and winter, some homes in Princeville, Tarboro, and other hard-hit cities were gutted for reconstruction or torn down. As federal money and low-interest loans trickled in, damp and crumbling sheetrock was pulled out, electrical wiring was replaced, and homes were put back into livable condition. Crews of volunteers from around the country were organized to help with the construction. Church groups, Habitat for Humanity volunteers, college students, and reassigned state employees rolled up their shirt sleeves and went to work rebuilding homes. In many areas, the federal money went toward buyouts in which homes and land in the most flood-prone areas were purchased and residences were torn down. Federal Emergency Management representatives estimated that as many as 10,000 flooded properties could ultimately be

bought, though only about half that many were purchased in the first year after the storm. The buyouts removed thousands of families from high-hazard areas, protecting them from the risks of future floods. The result was that some neighborhoods became ghost towns, with only one or two occupied homes in a community that once had fifty or more.

Though the economic impact of hurricane Floyd in North Carolina was staggering, the heartbreaking loss of 52 lives in the state was the disaster's greatest tragedy. No other hurricane of the twentieth century had killed as many North Carolinians; in all of the state's recorded history, only an un-named hurricane in 1883 had killed more (53). The report of 52 deaths came from the state medical examiner, though the National Hurricane Center's count was 56 direct deaths in the United States, including 35 in North Carolina. More than a dozen lives were lost in other states, including 6 in Pennsylvania, 6 in New Jersey, 3 in Virginia, 2 each in Delaware and New York, and 1 each in Connecticut and Vermont. Some media reports suggested that Floyd caused as many as 77 fatalities across the country. But no matter whose fatality figures are used, Floyd's death toll was still the largest of any U.S. hurricane since Agnes swept from Florida to the Northeast in 1972, killing 122.

In North Carolina, reports of fatalities began to trickle in during the first hours after the storm made landfall. Then, as the flooding worsened and roads became lakes, frustrated drivers fell prey to swiftly moving currents, and the list of storm victims grew longer. Among all the painful lessons learned from the Floyd disaster, none was more critical than understanding the deadly consequences of driving through floodwaters. Well over half of the state's storm-related fatalities happened when motorists attempted to drive over submerged roads.

Most of the deaths occurred within the first few days after Floyd made landfall, but because the high waters lingered, reports of fatalities continued to emerge weeks after the storm. Understandably, most of the dead were victims of freshwater flooding in inland areas. In a summary report on Floyd, the Hurricane Center noted that 86 percent of the deaths were due to inland flooding, 55 percent were vehicle related, and 80 percent were male. These figures follow a trend that has emerged in recent decades, wherein most hurricane deaths are caused by freshwater flooding away from the coast—a far cry from earlier decades, when powerful storms drove storm surges into unprepared coastal communities, killing hundreds.

The following list of the fatalities in North Carolina comes from a report in the *News and Observer*, with additional information from the state medical examiner:

Eulalia Mills Aldridge, 87, of Spring Hope, drowned when the car she was riding in was swept off a U.S. 64 overpass.

Sherry Boyer, 54, of Pennsylvania, died of a heart attack when her van was struck with water from a burst dam on Interstate 95.

Chris Brown, 75, of Rocky Mount, drowned when he fell into a ditch of rushing water in Edgecombe County.

Paul Buco, 70, of Wilmington, died when floodwaters swept his car off Interstate 40 in Pender County.

Badger Chandler, 76, of Vanceboro, died when his car sank in deep water on N.C. 43.

Aaron Child, 18, of Leland, was found drowned by campus police in a stream near East Carolina University.

Ransom Cole, 70, of Wallace, died when he was swept away by floodwaters near his home.

Judy Core, 37, of Goldsboro, died fleeing a fire in her home that was ignited by a kerosene lamp.

Brandon Davis, 8, of Maryland, died when a car driven by his father tried to get around a washed-out bridge.

Kenneth Denning, 51, of Mount Olive, died when his pickup truck was swept into floodwaters on Steven Mills Road.

James Driver, 50, of Wilson. Details unknown.

Destiny Flowers, 3, and Cabrina Flowers, 5, of Pinetops, died when the small boat in which they were being rescued capsized.

Mario Gomez, 26, and Silverio Beltran Gomez, 35, of Grimesland, drowned when their car was swept away in floodwaters on Blackjack-Grimesland Road.

Linwood Gooding, 54, of Kinston, drowned after he abandoned his stalled car in floodwaters on N.C. 11.

Randolph Grandberry, 38, of Harrellsville, died of an apparent heart attack while bailing out his girlfriend's car.

Benjamin Harrison, of Nash County, was the last fatality of the storm to be recorded. His body was found in a quarry in November 1999.

Lou Hendricks, 55, of Norlina, died when floodwaters swept the car she was riding in off S.R. 1600 in Warren County.

Gerald Hoke, 63, of Orange County, died of exposure. His body was found after the storm.

George Jefferson, 43, of Windsor, drowned when his pickup truck ran off a road.

Ossie Lee Jenkins, 65, of Whiteville, died of a heart attack after being rescued from his car.

Ernita Edwards Jones, 22, of Greenville, was killed when she lost control of her car on N.C. 42 and struck an oncoming car.

William Teague Jones, 31, of Hot Springs, was working as a contractor for Carolina Power and Light in New Hanover County when he was electrocuted.

Reginald Jones, 37, of Maxton, died when the car he was riding in hit a fallen tree.

Osseynna Jones, 1, of Robeson County, died of burns suffered in a fire at her home.

George Jones, 72, of Warrenton, died with Lou Hendricks when the car they were riding in was swept away in Warren County.

Eusebio Maldonado, 30, of Clayton, died after water swept his vehicle into Little River Creek in Johnston County.

Ben Mayo, 50, and Vivian Mayo, 45, of Pinetops, died after a small boat they were using to evacuate flooded houses capsized.

Keisha Mayo, 24, of Zebulon, died in the same boat.

David Mills, 79, of Spring Hope, drowned with Eulalia Aldridge when their pickup truck was washed off a U.S. 64 overpass.

Paul Mobley, 31, and Emily Mobley, 5, of Clayton, drowned when their car was swept into a creek on a back road in Johnston County.

Marvin Moody, 43, of Maxton, died when the car he was riding in ran into a fallen tree on a road near Pembroke.

Reiford Nichols, 55, of Grifton, drowned in a boating accident on the Neuse River on his way to pick up his son.

William Nixon, 47, of Currie, drowned after floodwaters swept his truck off a back road north of Burgaw.

Leon Penland, 22, of Hayesville, a National Guard military policeman, died when the Humvee he was driving overturned on Caswell Beach.

Richard Phillips, 40, of Nashville, died after his car plunged into floodwaters on Nashville Road.

Mitchell Piner, 42, of Wallace, died after his truck was pushed from N.C. 41 into rushing water.

Charlotte Poythress, 47, of Gaston, drowned trying to get to safety after her fiancé's van stalled on a flooded roadway near Fishing Creek at the Nash County Line.

Otis Reid, 51, of Pinetops, drowned when his mobile home was flooded during the storm.

Ronald Russell, 43, of Greenville, drowned after abandoning his car on a flooded road.

Ather Smallwood, 96, of Windsor, died of an apparent heart attack after being returned to the rest home where she was living before the evacuation.

Roger Smith, 86, of Grifton, died of complications from a broken hip incurred during the hurricane.

James Stokes, 65, of Hobgood, was found dead in his car south of Scotland Neck.

Larry Summerlin, 63, of Mount Olive, drowned on his way to work when his vehicle was swept off N.C. 11.

Teshika Vines, 5, of Pinetops, died after a boat used to escape flooded houses capsized.

Artemus Westry, 46, of Nashville, died when his car was washed off U.S. 64 near Stony Creek in Nash County.

Cheryl Whitley, 42, of Goldsboro, drowned when her car washed off a two-lane road near Stoney Creek Bridge.

James Wilder, 60, of Trenton, drowned when he was swept into floodwaters off N.C. 58 near Trenton. Wilder was a N.C. Department of Transportation worker who died while trying to help a stranded motorist.

James Wilson Jr., 33, of Godwin, died when the vehicle he was riding in struck a fallen tree on a road south of Dunn.

It might normally be expected that a particular state or region of the country would experience a true benchmark hurricane only once in a generation. But for thousands of North Carolinians, especially those whose homes were destroyed, two major hurricane disasters within three years was a perplexing and exhausting experience. Though both hurricanes Fran and Floyd were truly memorable events in Tar Heel history, each had its own unique recipe for destruction and misery. Floyd's impact on the state, as illustrated in the following statistical summary, establishes it without question as North Carolina's single greatest disaster.

Floyd statistics for North Carolina:

66 counties declared disaster areas

Total damages between $5.5 and $6 billion

52 deaths reported by the state medical examiner

63,000 houses flooded

7,300 homes destroyed

86,954 people registered with FEMA for aid in the first five months after the storm, compared with 72,730 who sought assistance after Fran

U.S. Small Business Administration approved more than 12,600 low interest loans

FEMA paid 12,830 claims totaling $141 million in National Flood Insurance, though less than 13 percent of homes in the affected areas were covered

Over $19 million raised from private sources for the Hurricane Floyd Relief Fund

Agricultural damages included 50 percent of the sweet potato crop, 40 percent of the cotton crop, and $92 million in tobacco losses

Among agricultural losses were more than 30,000 hogs, 2.4 million chickens, and 700,000 turkeys

At the peak of the crisis, 235 Red Cross shelters were open, housing a population of almost 50,000; over 1.3 million hot meals were served

1,400 roads and highways closed because of high water

Between 50,000 and 75,000 cars damaged by floodwaters

40 dams failed; another 61 sustained damage

At least 50 hog lagoons and 24 municipal sewage plants submerged

More than 1.5 million customers without electricity

More than 1,500 people rescued from flooded areas

NOR'EASTERS

A young boy scrambles to avoid a wave breaking over a high dune in Atlantic Beach during the Storm of the Century in March 1993. (Photo courtesy of Scott Taylor)

Many northeasters have affected the North Carolina coast over the years, causing hurricane-like destruction. A March storm in 1926 flooded Front Street in Beaufort, prompting the evacuation of many waterfront businesses, including the Ramsey Grocery Co. (Photo courtesy of the Carteret County Historical Society)

The Ash Wednesday Storm of 1962 was as potent as many severe hurricanes. More than eighteen hundred houses were destroyed along the mid-Atlantic coast. (Photo by Aycock Brown; courtesy of the Outer Banks History Center)

The press labeled it "The Storm of the Century." Residents along the eastern seaboard called it "the hurricane with snow." But the great storm of March 1993 wasn't a hurricane at all, at least not by definition. It was, instead, a powerful winter storm that meteorologists refer to as an extratropical cyclone. Old-timers will tell you it was just a mean nor'easter.

Extratropical cyclones have many of the same characteristics as hurricanes. Strong counterclockwise winds rotate around an area of low pressure, spreading their destructive energies over thousands of square miles. While over the ocean, these winds pile water along the shore, creating a surge effect that floods low-lying areas and brings extensive beach erosion. They can also deliver heavy rains or, as in the case of the '93 storm, blizzardlike snowfall.

Extratropicals differ from hurricanes in that they lack a central warm-air mass and a well-defined eye. They typically occur during the winter months and may even originate over land. Some begin when a strong low-pressure system in the upper atmosphere transfers its energy to a developing low-pressure system off the mid-Atlantic coast. Others form near the Gulf of Mexico, cross into the South Atlantic, and drift into position off Cape Hatteras. These systems are known as "Hatteras Lows" and have been responsible for billions of dollars in losses along the Atlantic coast over the years.

Northeasters are named, of course, for the ceaseless northeast winds that batter the coast as the storm edges by. And that can sometimes be days, as extratropicals are notorious for stalling offshore while their wind-driven waves

pound the beaches. Erosion is often severe during these winter storms, frequently cutting away at protective dunes and undermining beachfront decks and walkways. Over the years, northeasters have taken a heavy toll on the North Carolina coast, and some have been even more destructive and costly than hurricanes.

The March Superstorm of 1993 was one extreme example of the extratropical phenomenon. This unnamed cyclone carved a deadly path from Cuba, where 3 died, to the Canadian Maritimes, where 4 more lives were lost. The total number of deaths was 238, not including the nearly 50 sailors lost from vessels that sank off Nova Scotia and in the Gulf of Mexico. In all, $1 billion in damage was reported in the United States.

This monster storm was known to most in the east as the Blizzard of '93. It delivered record cold to the Deep South and dumped neck-deep snow from Georgia to Canada. Mount Mitchell, North Carolina, recorded fifty inches, and travel was stalled by the snow in numerous eastern states. The storm's rapid development caught many off guard, even though the National Weather Service had forecasted it almost perfectly. Several hikers were stranded in the Great Smoky Mountains but were later rescued after an extensive search.

In Florida, the storm spawned several tornadoes that wreaked havoc across the state. High winds made deadly projectiles out of debris still on the ground from the aftermath of hurricane Andrew, which had struck the same area the previous summer. In all, forty-four Florida residents were killed in the March storm, surpassed only by the fifty who died in Pennsylvania.

In North Carolina, much of the western half of the state was blanketed by the heavy layer of snow laid down by the storm. Along the coast and through-

When the floodwaters subsided following the Ash Wednesday Storm of 1962, sand was deep in the streets of Nags Head. (Photo by Aycock Brown; courtesy of the Outer Banks History Center)

Strong northeast winds pushed waves across the Outer Banks during the Halloween storm of 1991, flooding portions of Highway 12. (Photo by Drew Wilson; courtesy of the Outer Banks History Center)

Highway 12, the only roadway from Ocracoke to Nags Head, is prone to flooding and washouts during hurricanes and northeasters. (Photo by Drew Wilson; courtesy of the Outer Banks History Center)

Sea foam spattered this Buxton home during a strong northeaster in February 1973. (Photo by Ray Couch; courtesy of the Outer Banks History Center)

A winter northeaster claimed this oceanfront home in Nags Head. Northeasters and hurricanes in recent years have altered the shoreline in the Nags Head area, causing several houses to be lost to the Atlantic. (Photo by Drew Wilson; courtesy of the Outer Banks History Center)

Each year winter storms bring changes to North Carolina's beaches by adding sand to some areas and eroding it from others. Oceanfront property owners are often forced to adapt to the reshaping of their "front yards." (Photo by Drew Wilson; courtesy of the Coastland Times)

out many eastern counties, gusting winds of hurricane force whipped trees and structures for more than eight hours. In Southport, Wrightsville Beach, Morehead City, and Manteo, sustained winds between 70 and 90 mph were reported, with gusts surpassing 100 mph.

Because of the inland track of the storm, very little precipitation fell along the coast. Winds from the south and west lifted salt spray off the ocean and deposited it on vegetation, power lines, and electrical insulators. As a result, power was disrupted for more than twenty-four hours in many areas, at a time when nighttime temperatures dipped into the twenties. The heavy salt layer "burned" the leaves of trees and shrubs for miles inland. Governor Jim Hunt declared forty North Carolina counties disaster areas in the wake of one of the worst winter storms in years.

One of the most severe northeasters in North Carolina history was the infamous Ash Wednesday Storm of 1962. This intense nonhurricane pounded more than five hundred miles of the mid-Atlantic coast from March 7 to 9. For sixty hours, fierce winds and raging surf battered the Tar Heel coast, especially the northern Outer Banks. The power of the storm and the destruction it left behind cause it to be ranked alongside the worst of North Carolina's hurricanes.

Well-known author David Stick chronicled the disaster in his book *The Ash Wednesday Storm*, which features dozens of photographs by Aycock Brown. Stick detailed the destruction caused by the storm's excessive floods—floods made worse by the alignment of the sun, moon, and earth that created one of the highest lunar tides of the year. Surging ocean waters flattened the protective dune line from Kill Devil Hills to the Virginia border. Near-record storm tides at Cape Hatteras opened an inlet two hundred feet wide just north of Buxton, which was later filled in by the U.S. Army Corps of Engineers.

The Ash Wednesday Storm was more brutal to oceanfront property than many hurricanes. Thousands of homes and cottages were severely damaged, and about eighteen hundred dwellings were destroyed. Total structural damage to the mid-Atlantic states was estimated at $234 million. Although the storm wasn't exactly a typical winter northeaster, it served as a reminder to coastal residents that hurricane-like winds and floods can strike during any season of the year.

CREATURES IN THE STORM

For centuries, hurricanes have lashed the North Carolina coast, battering the people and property of the state. Countless stories have been told of the awesome forces of wind and water and of the harrowing ordeals faced by the victims of these storms. Among these stories of human survival, however, are numerous accounts of how the state's nonhuman creatures survived. Domestic animals and wildlife are, in many ways, just as vulnerable to hurricanes as humans. Their stories have also become part of our fascination with the hurricane phenomenon.

Through the years many isolated portions of the Tar Heel coast have provided haven for feral horses. These hardy creatures have endured blistering summers and barren winters on the thin strips of barrier beach that line the shore, but the ravages of severe hurricanes have thinned their numbers and caused mass die-offs on several occasions. The San Ciriaco hurricane of 1899 drowned hundreds of "banker ponies" as well as scores of cows and goats. So many were reportedly lost that burning their carcasses became the only effective means of eliminating a potential health hazard on the banks.

During hurricane Fran, thousands of trees toppled into pastures across the state, sometimes striking and killing horses and livestock. But some trees proved even more lethal in the days following the storm, when horses that munched on the wilted leaves of fallen red maple, cherry, or elderberry branches became sick and died. The staff of North Carolina State University's College of Veterinary Science scrambled to save what animals they could and issued warnings to horse owners to remove the downed trees as quickly as possible. Cherry tree leaves contain cyanide, but the leaves of all three species are highly toxic to horses and are usually lethal.

In the hurricane of 1933, some down-east residents reportedly brought their domestic stock inside their homes as floodwaters approached. Pigs, goats, chickens, and even cows were coaxed up to second-floor rooms to escape the rising tide. Many farm animals were not so fortunate, however. Most of those left to fend for themselves either drowned or were scattered by the storm.

One peculiar story appeared in the *Beaufort News* one week after the storm of '33:

Down at Roe, which is located on the north end of Cedar Island, some men spied a forty-pound shoat [a young hog] lodged in the crotch of a tree Monday morning after the storm about fifteen feet from the ground. This animal had evidently been carried to this place by the high tide and terrific hurricane wind Friday night. In order to prevent the apparently dead pig from decomposing and causing both stench and disease, it was decided that the animal should be removed and buried.

One of the men climbed the tree with a saw and started to remove one of the limbs of the crotch, so that the shoat would fall to the ground. About the second or third stroke of the saw, the pig came to life and let out an unearthly and demon-like squeal that echoed and reached through the woods around Roe. This unexpected resumption of life on the part of the supposedly dead shoat pretty nearly frightened the rescuer to death. After it dawned on the bewildered men that the pig was really alive, they quickly removed him from the tree crotch.

Well-known "fish-house liar" Rodney Kemp tells that in the late nineteenth century, residents from the island communities of Portsmouth and Diamond City would often flee by boat upon the approach of a severe storm. They would carry a few possessions with them while they visited friends and relatives on the mainland. Prior to these excursions across the sound it was common practice to gather all the chickens and tie their legs together with twine. The islanders would then place the birds on their backs in the bottom of their boats for the journey to the mainland. Kemp says that the chickens became so familiar with this procedure that they would instinctively jump into the boats with their legs in the air upon the first signs of a storm.

Waterfowl can sometimes become victims of severe hurricanes. Hunters report that ducks seem to "disappear" after these storms, perhaps seeking cover in areas farther inland. In the book *Reflections of the Outer Banks* by Donald and Carol McAdoo, former Corolla postmaster Johnny Austin described the effects of high winds on some waterfowl: "The ducks and geese used to be so plentiful around here that when a storm came up it wouldn't be unusual to pick up $15 or $20 worth that had killed themselves flying into that top wire of the lighthouse. And you only got five or six dollars a barrel for them, depending on the kind they were, so it took a lot to mount up to $15."

As hurricane Donna tracked toward North Carolina in 1960, a large flock of seagulls became trapped inside the storm's eye. Military radar confirmed the presence of the birds, which were carried hundreds of miles northward by the storm. Many of the flock finally escaped near Wilmington when the eye distended to sixty miles in diameter, but hundreds of dead gulls washed up on the beaches between Carolina Beach and Topsail Island in the days following the hurricane.

After hurricane Fran, bird watchers near Raleigh were treated to a rare sight—hundreds of seabirds were spotted around the Triangle. As the storm pushed across the state, it carried with it a collection of birds usually found over the Gulf Stream, including petrels and an Audubon shearwater. Most of the inland sightings were near Jordan and Falls Lakes. Seabirds usually fly ahead of a hurricane but sometimes travel within the eye as it moves over land. Often they drop out over large lakes, then rest up before their return to the Atlantic. Wildlife experts reported that the variety and quantity of coastal and seabirds found at Jordan Lake on the Saturday after the storm was unprecedented.

Hurricane floods often flush out a potential hazard for humans: venomous snakes. During the Sea Islands Hurricane of 1893, numerous deaths were reported from snakebites in South Carolina. Rattlesnakes and cottonmouth moccasins were the likely culprits. In one hurricane, a family was forced to climb onto an oak to avoid the rapidly rising tide, only to find "the branches filled with copperheads and other serpents." Snakes have sometimes escaped the floods by finding refuge in homes and furnishings. After the hurricane of 1876, one Hyde County man found a rattlesnake coiled inside his dresser drawer. The reptile was likely seeking a dry location to ride out the storm.

Many stories have been told of the hurricane's effects on creatures of the sea. After the August hurricane of 1881, the army's chief signal officer for Carteret County wrote in his annual report:

> Morehead City, N.C., 24th, over thirty hours in advance of the storm, the skies became blackened with seabirds of every kind, size, color and description, moving rapidly towards the west, as if fleeing from the violence of the coming storm. The strange conduct of the birds was equaled, if not surpassed by the finny tribes, as shown by the latter's rapid flight up Newport River, a narrow, turbid stream. All through the day the fish, in schools of millions, passed up the stream, followed by great droves of porpoise, so thick that the river looked like a slowly moving stream of ink.
>
> 27th, birds slowly returning, at Newport, where the stream is very narrow, the fishes and porpoise were so wedged in that they could not move either up or down. The above incident would appear to give evidence of the possession of a wonderful instinct by birds and fishes.

Surging tides sometimes retreat as quickly as they advance. After hurricane Donna passed through Nags Head, "flopping bass and other live fish" were picked up off the streets. After Hazel, a large flounder was found on a sidewalk in Carolina Beach, and blue crabs were seen on the streets of Morehead City. On more than one occasion, the people of Ocracoke have found fish in their furniture, as storm tides washed through their living quarters and deposited marine life all about.

An unusual fish story came out of Raleigh in the days following hurricane Fran. According to the *News and Observer*, Faye and Paul McArthur's house was bashed by falling trees, as were many of their neighbors' homes on Beechridge Road. A large pine fell across their backyard pond, which was home to their colorful collection of 16-to-24-inch-long koi. On the morning after the storm, the McArthurs ventured out to inspect the damage and encountered two young boys, who asked them, "You want to see the biggest goldfish you ever saw?" After following the boys across Glenwood Avenue, Faye McArthur reported: "It was one of mine. My Israeli koi. In a little puddle across the street. It must have flown 100 feet from my backyard, over the roof, across Glenwood."

In his book *Ocracokers*, author Alton Ballance relays a story told to him by his grandfather. During the San Ciriaco hurricane, witnesses reported seeing two porpoises swim through the village of Ocracoke when floodwaters inundated the island. For a short time, they became lodged in the forked branches of an oak tree, but rolling waves helped free the pair. They swam away and were last seen crossing the island and entering Pamlico Sound.

Carolina Beach resident Granger Soward spotted an unusual visitor at his Canal Drive home during the onslaught of hurricane Fran. For several hours, Soward and four friends watched the storm surge push into his yard, cover his mailbox, and float cars and debris around his neighborhood. But then, as the storm's eye passed over them, a strange omen appeared at the back door. A manatee drifted peacefully into the yard and lingered for about an hour before turning back to deeper water.

Fishermen have been known to land some unusual catches in the days following hurricanes. After the Great Atlantic Hurricane of 1944, large tuna and marlin were reportedly trapped in the surf along the Outer Banks. After hurricane Emily passed by Carteret County in 1993, one fishing pier reported several unusual landings. In addition to a 7-foot, 280-pound bull shark and a silver snapper (normally common to the Gulf Stream), a 27-inch sailfish was caught and released at Sportsman's Pier in Atlantic Beach. "It's been pretty strange," noted one angler. "Emily must have scared all the fish onto the beach."

Powerful hurricanes can be very disruptive to freshwater and marine fish populations. After hurricane Hugo in 1989, large fish kills were reported in the lakes and rivers of North and South Carolina. Most of the kills were the result of lowered oxygen levels in the water, caused by tremendous quantities of leaves, limbs, and other organic matter that flowed into streams and lakes. In some areas, fish kills resulted from wedges of salt water that were pushed upriver by abnormally high tides. Hurricane Fran produced its share of fish kills in North Carolina, especially in the Neuse and Cape Fear river basins. Not only were these rivers affected by tons of organic debris, but the flooding of nutrient-rich farmlands and hog lagoons, along with the discharge of millions of gallons of untreated human sewage from failed treatment plants, added to the kills. But not all the fish in these rivers perished. According to *North Carolina Wildlife*, biologist George Beckwith witnessed the escape of thousands of fish along the lower Neuse immediately after Fran. "What we saw at Oriental was a mass exodus of fish out of the Neuse River moving ahead of the dead water," Beckwith said.

Along the coast, hurricanes often decimate the nests of loggerhead sea turtles, which are laid in shallow sands on the upper edges of North Carolina beaches. Loggerheads, protected as a federally threatened species, typically lay their nests during the warm summer months, just before the most active portion of the hurricane season. The high waves and heavy erosion from hurricanes Bertha and Fran were particularly destructive to the turtle population, wiping out at least 230 nests along North Carolina's beaches. Hardest hit were beaches on the central and southern coasts, including Topsail Island, where 96 nests had been reported in the previous year.

Shellfish populations can be affected by hurricanes as well. North Carolina's harvests of white and brown shrimp have been affected by many storms through the years. Surging tides and excessive rains can prematurely wash shrimp out the inlets and into the ocean. And heavy rainfall can produce bacterial runoff that can force the closing of shellfish waters along the coast. After most hurricanes, the North Carolina Division of Marine Fisheries surveys the impact the storms may have had on commercial species.

Wildlife officials note that although some animals may be affected by a severe hurricane, most manage to endure these storms without harm. According to South Carolina wildlife experts, Hugo offered an extreme test of this theory. Large animals, such as deer, suffered few losses during the storm and actually may have benefited from increased browse created by shrubs and herbs that flourished where trees were knocked down. Entire forests may have been lost, but the animals within them survived fairly well. Experts agree that wildlife can endure a natural disaster like a hurricane far better than they can survive an oil spill or some other man-made catastrophe.

Domesticated animals, however, do not always fare as well. In the hours and days following hurricane Floyd, thousands of people across eastern North Carolina fled their homes for local shelters or became trapped by raging floodwaters. Some were rescued from their rooftops and others escaped by boat. Often left behind in the lingering floods were their family pets. Dogs, cats, birds, hamsters, and other animals became unwilling victims of the disaster, as they could not join their owners in area shelters. Many dogs and cats became lost and disoriented; others drowned. But a large number were rescued by special teams that worked the neighborhoods by boat. A Denver group from the American Humane Association, funded by the Animal Planet cable channel, was among the rescue teams. Ben Drotar of Animal Planet told the *News and Observer* of their initial success: "The first day, we rescued sixteen dogs, six horses, one cat, and a duck." Farm animals were also among the stranded, and many were brought to safety during some rather unique rescue missions, including several horses and cows that were lifted out by helicopter.

More than one thousand dogs and cats were rescued from floodwaters within the first week following Floyd. Many were returned to their families, but in the continuing chaos of the disaster, some animals could not be matched with their owners. Many of these were sick or injured and needed immediate treatment. Teams of veterinarians converged on N.C. State University in Raleigh, and an animal field hospital was set up in former storage space behind the North Carolina Museum of Art. At one time more than eighty cats and dogs were housed in the facility, where they received antibiotics and were treated for their injuries.

One Jones County resident was surprised by what waited for him when he returned to his home after Floyd's floods subsided. Dennis Schafer told *WRAL News* that he found six snakes coiled in his kitchen sink. Driven from their homes by the storm's rising floodwaters, the reptiles ended up in Schafer's kitchen in their search for high ground. Though he wasn't bitten, Schafer reported that the snakes "hissed at him when he tried to shoo them out."

Many who have survived the frightening passage of a hurricane have done so with man's best friend at their sides. During hurricane Fran, one woman reported that her dog instinctively whimpered and barked, waking her from sleep just moments before a tree crashed through the roof and crushed her bed. During another storm, rescuers located a frail and elderly man trapped in a rising flood by zeroing in on the bark of his faithful companion.

Sometimes, however, dogs have been known to even save themselves. Al's Auto Salvage in New Bern was guarded by Petey, a junkyard dog all of ten inches tall. Skip Crayton, owner of the shop, gave him that name because of his resemblance to the mascot on the television show *Little Rascals*. On the Friday night after Fran roared through the state in 1996, flooding inside the auto salvage building rose sixteen inches above the floor. The following morning, when Crayton arrived at the shop, he opened the door and out came Petey—covered up to his neck with mud and oil. Crayton reasoned that the dog must have swam for six to eight hours continuously, keeping his head just above water to stay alive. Petey slept for two days.

THE NEXT GREAT STORM

(Page 277)
How will North Carolina's coastal beaches and inland communities endure the next great storm? (Photo by the author, taken near his home in Pine Knoll Shores after hurricane Fran in 1996)

As hurricane Floyd swept past Florida and turned toward the Carolinas near the peak of the 1999 hurricane season, there was actually some cause for relief on the part of local officials and storm watchers along the Tar Heel coast. Though the storm clearly was headed their way, most observers were thankful that the once-mighty hurricane was weakening as it edged northward. Just days before it had been a strong category four, and fears of a Hazel- or Hugo-like catastrophe seemed to wane with each update from the National Hurricane Center. Floyd was coming ashore with heavy rains and destructive tides, but at least it was a weaker system—as many viewed it; it would make landfall as *only* a category two.

What occurred next, of course, resulted in the greatest disaster in North Carolina history. Who could have predicted that such a thing would happen? Who could have foreseen that anything less than a *major* hurricane would cause such destruction?

A study of the many storms that have affected the region in the past yields one key observation: though every hurricane is different in size, intensity, and orientation, each has its own unique potential for disaster. Floyd may not have buffeted the state with fearsome winds, but its drenching rains fell on a supersaturated coastal plain and produced record flooding. And like those of hurricane Agnes in 1972 and tropical storm Alberto in 1994, Floyd's inland floods proved deadly, destructive, and difficult to anticipate. Because most of the prelandfall hype had focused on the incredible winds and potential tides, state and local officials were slow to grasp the time-delayed impact of the disaster.

Barrier island resort towns and other coastal communities have long been sites of emphasis for hurricane preparedness. Their evacuation plans are well scripted, their building standards are high, and their sense of hurricane-readiness is keenly focused. This is only logical, since we have come to expect that most hurricanes will have their greatest impacts on coastal areas near the point of landfall. But more and more, our nation is experiencing hurricanes whose winds and rains cause catastrophic destruction in communities many miles from the coast. Though their effects on beaches should not be downplayed, North Carolina's three costliest storms—Hugo, Fran, and Floyd—all visited chaos on noncoastal cities like Charlotte, Raleigh, and Greenville. Though inland communities have not traditionally adhered to the same level of preparation as those on the coast, things are changing. After suffering through the floods of Fran and Floyd, dozens of towns far from the ocean are now using their disaster experiences to forge new plans and prepare for future events.

But even if those inland areas had been better prepared for a hurricane, the scope of the Floyd disaster would still have been more than they could handle. The storm's unprecedented impact caused many to question why the

flooding was as bad as it was. Could forecasters not have foreseen the magnitude of the event? Could it happen again? What should we expect when the next great storm hits North Carolina?

Though the National Weather Service knew Floyd would cause flooding if it tracked over eastern North Carolina, no one expected the massive innundation that occurred. Forecast models predicted record flooding for many places with only a projected eight to ten inches of rain, yet the actual rainfall was almost twice that in some areas. Still, the Weather Service did its job by issuing emphatic warnings well in advance of the storm. Steve Harned, director of the National Weather Service office in Raleigh, told the Associated Press: "The rains from Floyd were so far beyond anything we have ever seen that it really couldn't be forecasted with any degree of accuracy. We don't use the word 'biblical,' but we probably came as close as we ever will with Floyd. They were the most strongly worded flood watches and warnings we have ever written."

After the storm, scientists and government officials searched for clues to explain the breadth and severity of the flooding. Though media reports originally described the disaster as a "500-year flood," officials at the U.S. Geological Survey later said that Floyd was perhaps a 150- or 200-year event. The total

Those communities hardest hit by Fran struggled to recover and rebuild in the months following the storm. By the spring of the following year, some homes on Topsail Island remained just as the hurricane had left them. (Photo by Chuck Liddy; courtesy of the Raleigh News and Observer

The streets of Wrightsville Beach will again be flooded by a severe hurricane, just as they were during Fran in 1996 and Hazel (above) in 1954. (Photo courtesy of the News and Observer Publishing Co./N.C. Division of Archives and History)

Through the years, hurricanes have cost our nation many billions of dollars and thousands of lives. Recent hurricanes have caused fewer deaths, but property damages continue to exact a high cost. The next great hurricane to strike the North Carolina coast will likely fit that same pattern. With advance preparation and adequate warning, though, the death toll can be minimized. (Photo by Aycock Brown; courtesy of the Outer Banks History Center)

Hurricanes and northeasters are constantly reshaping the geography of the North Carolina coastline. Structures built on the exposed portions of our barrier islands remain vulnerable to the changes these storms might bring. (Photo by Aycock Brown; courtesy of the Outer Banks History Center)

amount of rain was substantial, but not as rare as was first believed. Soon others were suggesting that the real cause of the disaster was not "an act of God" but was, instead, the landscape-altering activities of man. As USGS hydrologist Jerad Bales told the Associated Press: "We know it was a big event, but it won't be as unusual as many first thought it was. All we can do is make statistical estimates. The landscape is changing. There are buildings, bridges, culverts, and lots of development in the watershed. It all has an effect. And so really we don't know for sure how big or unusual it was."

In a poststorm editorial in the *News and Observer*, East Carolina University geologist Stan Riggs described the problem of urbanization in the Piedmont and upper Coastal Plain and how it contributed to the Floyd disaster. He asserted that "we created our own crisis in Eastern North Carolina through systematic and traumatic modification of our watersheds." Decades of growth have converted forests to farmland and farmland to shopping centers. Years of ditching, paving, bridge building, and stream straightening have altered the natural flow of water across the land. Construction of sewage treatment plants, junkyards, and swine lagoons has been allowed within the 100-year flood plain, creating added risks in times of major flooding. Marginal wetlands were ditched and drained. Roads were built with small culverts that effectively became dams during floods. In all, scientists and planners agree that these land-altering projects were largely to blame for the dramatic flooding seen during Floyd.

Flood maps, which are supposed to delineate areas of risk, were reviewed by

A long line of cars poured off of the Outer Banks in the hours before hurricane Emily struck in 1993. Because access to some North Carolina beaches is limited, evacuation plans must be implemented far in advance of approaching storms. (Photo courtesy of the Coastland Times)

a special state task force after hurricane Fran. The group's 1997 report concluded that, across the state, the existing maps were "inaccurate or grossly out of date." In the areas hardest hit by Floyd, the study found that 101 communities were using maps that were more than ten years old, some had maps nearly seventeen years old, and 70 towns did not have flood maps at all. Consequently, dozens of cities and counties throughout the east had for years allowed construction in flood-prone areas because of outdated maps. Had the maps been more current, more people would have been covered by flood insurance and the extent of the flooding would have been less surprising.

The cumulative impact of development can cause the floodplain to rise dramatically over a short period of time. When maps for Charlotte and Durham were updated, state officials found that the floodplain had risen an average of four feet above levels indicated on previous maps. Some structures built outside of the flood zone just a few years ago are now at risk of flooding in a 100-year storm. But urbanization is not the only human activity to influence change in the way floods occur. According to the North Carolina Division of Water Resources, land subsidence is another factor that has effectively raised the floodplain and made flooding worse. Subsidence is the sinking of land, in this case caused by decades of unsustainable withdrawals from underground aquifers. The division's Nat Wilson told the *Water Resources Research Institute News*: "I can document up to 9 inches of land subsidence in the Central Coastal Plain from 1935 to 1979. If you continue those rates of subsidence, then there has been up to 12 inches of subsidence in this area. A secondary rea-

As powerful as hurricane Andrew but almost three times its size, hurricane Floyd demanded the full attention of residents along the entire U.S. East Coast when it approached Florida on September 14, 1999. (Photo courtesy of NOAA)

Hurricane Floyd
NOAA-15 AVHRR HRPT
Multi-spectral False Color Image
September 14, 1999 @ 1244 UTC

son for the Floyd flooding is the reduced land surface elevations. A flood event will affect a much larger area because the land has subsided."

How North Carolina rebuilds after Floyd will have a profound influence on what happens in the next great hurricane. Finding the $20 million required to update flood maps was only one of the challenges lawmakers faced in developing policies to reduce the impact of future storms. Controlling where and how damaged properties are rebuilt was another critical issue. After the storm, Governor Hunt's administration proposed a new minimum building standard dictating that new homes would have to be built at least two feet above the 100-year floodplain. After pressure by cities, counties, and homebuilders, that proposal was scaled back to a one-foot minimum standard. Though some cities already mandated the one-foot standard, the new rule was designed to guide construction in the dozens of communities where previous restrictions had been few, thereby reducing losses in future floods.

But one of the best plans to minimize future property losses was the continuation of a program begun in the aftermath of hurricane Fran that involved a systematic retreat from low-lying areas along North Carolina's rivers and streams. Federally sponsored buyouts of at-risk properties in such locations should dramatically reduce damages in future floods. According to the *News and Observer*: "The buyout aims to be the largest in U.S. history, reclaiming the flood plains from more than 10,000 homes, swine farms, sewage plants, hazardous-waste sites, and drinking water systems. Officials hope forests and wetlands, nature's best flood control, will take their place." In the months following Floyd, thousands of homes and business properties across the eastern counties were purchased, later to be relocated or demolished. Eric Tolbert, the state's emergency management director, described the benefits of the program: "In essence, those are 4,000 families that will not be exposed to the consequences of another flood in eastern North Carolina. Beyond those buyouts, hurricane Floyd has presented a real opportunity to examine decisions made in the past, and to some degree, set a new course for eastern North Carolina."

But controversy continues to boil over the question of rebuilding communities where the risks are high. Some wonder why taxpayers should have to support the repeated reconstruction of storm-damaged homes and businesses in areas frequented by hurricanes. Nowhere are these questions more insistent than on the coast, where the state's barrier beaches took the brunt of at least five hurricanes in the late 1990s. State and federal lawmakers are still wrestling with the issue, but most recognize the economic and social complexities involved. State representative David Redwine, a lifelong coastal resident who co-owns an insurance company, offered a realistic perspective on building at the coast in an article run by the Associated Press: "You have to balance private property rights with what's in the best interest of the public. . . . The bottom line is: As long as insurance is available, as long as the infrastructure is available, people are not going to move back from the ocean. Everyone wants a little piece of paradise."

Among issues of hurricane vulnerability and public policy, few topics generate as much controversy as that of beach nourishment. For years property owners, environmentalists, geologists, and engineers have quarreled over the merits of spending millions to pump sand onto erosion-prone barrier beaches. Some argue that the beaches are always moving and shifting and that efforts to artificially stabilize them with dredged-up sand are futile. Furthermore, they point out that these multimillion-dollar, tax-supported projects are at risk of being washed away in a few short years by the next round of hurricanes and winter storms. Their philosophy focuses on pulling structures back away from the vulnerable ocean dune line and limiting new construction in areas of severe erosion. Between 1989 and 1995, about 250 buildings in North Carolina were moved back from the advancing ocean and resettled either on the same lot or on other lots. Perhaps the most famous structural retreat came in the second quarter of 1999, when the Cape Hatteras lighthouse was moved approximately 1,500 feet away from the threatening Atlantic.

Others, however, believe that beach nourishment can be cost effective and that it is necessary to protect the state's considerable economic investment on the coast. Carolina and Wrightsville Beaches are the sites of two of North Carolina's oldest nourishment projects, funded mostly by the U.S. Army Corps of Engineers since the mid-1960s. Both endeavors were designed for long-term erosion and hurricane protection and feature high-elevation dunes and broad beaches. Though engineers consider Floyd's storm surge to have been a 75-year event, not a single building behind the fortified dunes of those two beaches was lost in the storm. Nonetheless, other coastal communities have struggled to develop their own sand replenishment projects. In Carteret County, voters in 1999 rejected a $30 million bond referendum that would have financed the nourishment of seventeen miles of strand from Atlantic Beach to Emerald Isle. That project would have been funded by local taxpayers with no federal support. But lingering questions over the fairness of the project's funding scheme and uncertainty as to its effectiveness killed the proposal.

Though the debate continues, at least one prominent researcher sees value in the use of nourishment projects on some beaches. Spencer Rogers, a coastal engineer with North Carolina Sea Grant, has studied the effects of recent hurricanes on coastal areas that have been the recipients of sand nourishment and concludes that such projects can be very effective. "The beach nourishment and dunes proved to do exactly what was intended—protect the communities from hurricane-related erosion. It is a treatment. Erosion won't go away. Rather, nourishment replaces sand that would be lost to natural causes around the shoreline. To be successful, a long-term commitment for adding sand every two to four years is usually required."

With memories of Fran and Floyd still fresh in their minds, prudent storm watchers are already planning for the next hurricane. Though it may be unpleasant to speculate on the subject, there is little doubt about it—sooner or later, another disastrous storm *will* come. Perhaps it will be another category three like Fran, a category four like Hazel, or even a series of weaker storms with drenching rains like Dennis and Floyd. No one knows when or where it will strike, but we do know that eventually it will blast ashore and cause massive destruction—perhaps even greater than that caused by previous storms. There is nothing anyone can do to alter that foreboding reality, but the real question is: Will we be ready?

Obviously, we will rely on forecasters at the National Hurricane Center to warn us when the next big storm turns toward the Carolina coast. Technological advances have given meteorologists better tools with which to analyze storm movements, and forecast accuracy has improved over the last few decades. The computer models used to predict hurricane landfalls have become more sophisticated, and other advances, such as the addition of a Gulf Stream IV jet for high-altitude reconnaissance, should enhance the Hurricane Center's capabilities. The center already has a well-earned reputation for issuing timely, lifesaving warnings about approaching storms. But for all the meteorological improvements, predicting exactly where and when a hurricane will strike, and how strong it will be when it does, remains an elusive goal. The

Hurricane Center's average error in a twenty-four-hour forecast is currently around 100 nautical miles, down from about 120 nautical miles in 1960. Neil Frank, director of the center from 1974 to 1987, summarized the problem thus: "While I was at the Hurricane Center, there was only a 10 percent improvement in forecast error in twenty-five years. At the same time, the coastal areas were just exploding in population. The increase in population is far surpassing any small improvement we are realizing in our ability to forecast hurricanes."

Though meteorologists are improving their techniques for determining where hurricanes might go, they still struggle with forecasts of hurricane intensity. Our burgeoning coastal population faces its greatest threat from a rapidly intensifying storm that accelerates toward the shore. This is the scenario that emergency planners fear most and that hurricane meteorologists have the toughest time forecasting. Consequently, forecasters must be cautious, and coastal residents should expect to sometimes suffer through large-scale evacuations for storms that may weaken before they make landfall. Under these conditions, the maxim for life on the coast is truly "better safe than sorry." Hurricane Center director Max Mayfield described his agency's progress: "I think we will see a continued, gradual improvement in the track forecasts. In my opinion, however, we are several years away from making significant improvements in intensity forecasting. We really want people to know that there is little skill in forecasting rapidly strengthening hurricanes."

In planning for the next great storm, the citizens of North Carolina have now had the greatest of teachers—firsthand experience. After enduring so many hurricanes in the late 1990s, including major disasters with Fran and Floyd, they can use the lessons learned and the mistakes made to better prepare for the next event. A wealth of experience was gained by emergency management officials, state and local governments, and others who worked through the recovery process. Written plans and procedures may be necessary, but the efforts to coordinate disaster recovery on an unprecedented scale provided countless unwritten learning opportunities. In the wakes of Fran and Floyd, new relationships were forged and cooperation was extended between public and private agencies, governments, businesses, churches, relief organizations, and families. The next time a big hurricane rolls through the state, many mistakes made in previous storms will be corrected, cooperative efforts should be strengthened, and all those involved will have the benefit of past experience.

According to Eric Tolbert, director of the North Carolina Division of Emergency Management, coordination and planning on the state level have improved significantly with each disaster experience: "Perhaps the most pointed improvement is the philosophical change of being proactive versus reactive in response and recovery efforts. Through strategic placement of operational facilities and transportation and commodity contracts, the state is positioned to provide immediate support to impacted communities, while also saving the state money. We now place well-trained teams of state and local officials in county emergency operations facilities, which provide resource support to counties severely impacted. With experience the state's response was improved, which constituents expect, just as they will expect from hurricane Floyd to the next event. Floyd presented new response challenges: swift-water rescue in eastern North Carolina; heavy humanitarian airlift operations; disaster mortuary operations; and animal rescue. These consequences are now factored into response planning for future disasters, thus improvements are expected—but we'll face another set of challenges with the next event."

Though lessons were learned and plans improved after Floyd, emergency managers and hurricane forecasters still fear the worst when the next monster hurricane looms. In the late 1990s, atmospheric scientist and hurricane prognosticator William Gray of Colorado State University gave emergency planners even more to be concerned about. Not only had he accurately forecasted a string of seasons with above-average hurricane activity, but he also made a long-term forecast that was not promising. Gray, along with other hurricane researchers, believes that we may have entered a period of increased storm frequency that could well last for decades. Gray's research, which includes such factors as rainfall in Africa's Sahel region and the effects of El Niño and La Niña on the tropical atmosphere over the Atlantic, has enabled him to produce an annual prediction for Atlantic hurricane frequency since the early 1980s. The evidence now suggests that North Carolina and the rest of the Atlantic basin may experience more frequent hurricanes in the coming years than have been normal in the last few decades. It would certainly seem that a trend has emerged, since the last five years of the 1990s produced more hurricanes in the Atlantic than any other five-year period on record. It is unknown exactly how long this active period may last, but estimates range from fifteen to twenty-five years. In an interview after Floyd, Hurricane Center director Max Mayfield added: "We are obviously in an active period. It would be very foolish to expect hurricanes to miraculously stop coming. The message from the Hurricane Center has been very consistent. Every individual, every family, every business, and every community needs to have a hurricane plan and have it in place now before a hurricane threatens."

No one can say exactly when North Carolina will be hit by the next great storm. Most people generally accept the risk, just as midwesterners accept tornadoes and Californians live with the threat of earthquakes. Vast improvements have been made in hurricane forecasting, warning, communications, and recovery, but North Carolina still remains vulnerable. The problems faced by the state today are rooted in geography, population, behavior, and economics. Coastal evacuations are still a primary concern, but now we must face additional challenges across the state. And as more people move into the state and the population ages, the complexity of evacuation and recovery efforts will increase. But common sense is the ultimate weapon against the hurricane threat, and well-prepared residents know that while lost property is replaceable, lives lost are lost forever. If a major hurricane threatens, their cars will be packed and their plans in place to escape the approaching storm. They know it's not a matter of *if*— sooner or later another big one *will* come.

HURRICANE SURVIVAL **12**

WHEN A HURRICANE THREATENS

KEEP YOUR RADIO OR TV ON...AND LISTEN TO LATEST WEATHER BUREAU ADVICE TO SAVE YOUR *LIFE* AND POSSESSIONS

North Carolina will again be visited by severe hurricanes. The key to surviving the next great storm will be preparation. The following information is provided as a public service by the North Carolina Division of Emergency Management as a guide for hurricane survival.

BEFORE A HURRICANE THREATENS

Know the elevation of your home above sea level. Get this information from local Emergency Management officials. Your nearest Weather Service office can supply flood-stage data for area streams and waterways. Find out if your home is subject to storm-surge (tidal) flooding.

Know the maximum storm surge that might occur. Information about the potential for inland flooding and storm surge is available through your local Emergency Management office.

Know the route to safety if you have to leave. Plan your escape route early. Check with Emergency Management for low points and the flooding history of your route. Check the number of hours it could take you to evacuate to a safe area during peak evacuation traffic.

Know the location of the nearest official shelter. Emergency Management can give you the location of the shelter nearest your home and explain what you should bring with you. Plan for your family's safety. Know how to contact family members if the need arises.

THE STORM TIDE
MAY BE A HURRICANE'S GREATEST KILLER

TAKE PRECAUTIONARY MEASURES PROMPTLY WHEN THE WEATHER BUREAU ISSUES
HURRICANE WARNINGS

How safe is your home? Near the seashore, plan to relocate during a hurricane emergency. If you live in a mobile home, always plan to relocate.

Inventory your property. A complete inventory of personal property will help in obtaining insurance settlements and/or tax deductions for losses. Inventory checklists can be obtained from your insurance representative. Don't trust your memory. Keep written descriptions and take pictures. Store these and other insurance papers in waterproof containers or in your safety deposit box.

Know what your insurance will cover. Review your insurance policies and your coverage to avoid misunderstandings later. Take advantage of flood insurance. Separate policies are needed for protection against wind and flood damage, something people frequently don't realize until too late.

WHEN A HURRICANE WATCH IS ISSUED

Monitor storm reports on radio and television. If considering moving to a shelter, make arrangements for all pets. Pets are not allowed in shelters. If evacuation has not already been recommended, consider leaving the area early to avoid long hours on limited evacuation routes.

Keep a radio with extra batteries. Your transistor radio will be your most useful information source. Have enough batteries for several days, as there may be no electricity.

Evacuate early! (Photo by Drew Wilson; courtesy of the Outer Banks History Center)

Keep flashlights, candles or lamps, and matches. Store matches in waterproof containers. Have enough lamp fuel for several days, and know how to use and store the fuel safely.

Keep a full tank of gasoline in your car. Never let your vehicle gas tank be less than half-full during hurricane season; fill up as soon as a hurricane watch is posted. Remember: When there is no electricity, gas pumps won't work.

Make sure you have some cash. Remember that automated teller machines also won't work without electricity.

Store nonperishable foods. Store packaged foods that can be prepared without cooking and require no refrigeration. There may be no electricity or gas.

Keep containers for drinking water. Have clean, air-tight containers to store sufficient drinking water for several days. Local water supplies could be interrupted or contaminated.

Store materials to protect your windows. Have shutters, plywood, or lumber on hand to nail over windows and doors. Masking tape may be used on small windows but does not always protect against flying shards.

Keep materials for emergency repairs. Your insurance policy may cover the cost of materials used in temporary repairs, so keep all receipts. These will also be helpful for possible tax deductions.

WHEN A HURRICANE WARNING IS ISSUED

Listen constantly to radio or television. Keep a log of hurricane position, intensity, and expected landfall. Discount rumors. Use your telephone sparingly.

Leave your mobile home immediately. Mobile homes are not safe in hurricane-force winds.

Prepare for high winds. Brace your garage door. Lower antennas. Garbage cans, awnings, loose garden tools, toys, and other loose objects can be deadly missiles. Anchor them securely or bring them indoors.

Cover windows and other large glass. Board up or shutter large windows securely. Tape exposed glass to minimize shattering. Draw drapes across large windows and doors to protect against flying glass if shattering does occur.

Secure your boat. Move boats on trailers close to your house and fill with water to weigh them down. Lash securely to trailer and use tie-downs to anchor the trailer to the ground or house. Check mooring lines of boats in the water, and then leave them.

Store valuables and important papers. Put irreplaceable documents in waterproof containers and store them in the highest possible spot. If you evacuate, be sure to take them with you.

Prepare for storm surge, tornadoes, and floods. These are the worst killers associated with a hurricane. In a tornado warning, seek inside dry shelter below ground level. If outside, move away at right angles to a tornado; if escape is impossible, lie flat in a ditch or low spot. Remember that the surge of ocean water plus flash flooding of streams and rivers due to torrential rains combine to make drowning the greatest cause of hurricane deaths.

Check your survival supplies once again.

IF YOU STAY AT HOME

Stay indoors. Remain in an inside room away from doors and windows. Don't go out in the brief calm during the passage of the eye of the storm. The lull sometimes ends suddenly as winds return from the opposite direction. Winds can increase to 75 mph or more within seconds.

Protect your property. Without taking any unnecessary risks, protect your property from damage. Temporary repairs can reduce your losses.

Keep a continuous communications watch. Keep your radio or television tuned for information from official sources. Unexpected changes can sometimes call for last-minute relocations.

Remain calm. Your ability to meet emergencies will help others.

IF YOU MUST EVACUATE

Know where you are going and leave early.

Be prepared for the shelter. Take blankets or sleeping bags, flashlights, special dietary foods, infant needs, and light-weight folding chairs. Register every person arriving with you at the shelter. Do not take pets, alcoholic beverages, or weapons of any kind to shelters. Be prepared to offer assistance to shelter workers if necessary, and stress to all family members their obligation to keep the shelter clean and sanitary.

Don't travel farther than necessary. Roads may be jammed. Don't let your stranded auto become your coffin. Never attempt to drive through water on a road. Water can be deeper than it appears, and water levels may rise very quickly. Most cars will float dangerously for at least a short while, but they can be swept away in floodwaters. Wade through floodwaters only if the water is not flowing rapidly and only in water no higher than the knees. If a car stalls in floodwaters, get out quickly and move to higher ground.

Lock windows and doors. Turn off your gas, water, and electricity. Check to see that you have done everything to protect your property from damage or loss.

Carry along survival supplies. These should include a first-aid kit, canned or dried provisions, a can opener, spoons, bottled water, warm protective clothing, medications and prescriptions, spare eyeglasses, and a hearing aid with extra batteries, if required.

Keep important papers with you at all times. These should include a driver's license or other identification, insurance policies, property inventories, special medical information, and maps to your destination.

Common sense is the best defense against the threat of a hurricane. Property can be replaced; lives cannot. (Photo courtesy of the Coastland Times)

AFTER THE HURRICANE

If you are evacuated, delay your return until recommended or authorized by local officials.

Beware of outdoor hazards. Watch out for loose or dangling power lines, and report them immediately to local officials. Many lives are lost to electrocution. Walk or drive cautiously, as debris-filled streets are dangerous. Snakes and poisonous insects may be a hazard.

Do not drive on flooded roads, and avoid the temptation to go sight-seeing. Many fatalities occur because drivers attempt to drive on flooded roads. It is easy to underestimate water depth and current speed when a road is covered by flooding. Washouts may weaken pavement and bridges, which could collapse under your vehicle's weight. Never drive around highway barricades. Likewise, you should resist the temptation to sight-see; your tour could be dangerous and could hamper recovery efforts.

Guard against spoiled food. Food may spoil if refrigerator power is off for more than a few hours. Freezers will keep foods several days if doors are not opened after power failure, but do not refreeze food once it begins to thaw.

Do not use water until it is safe. Use your emergency water supply or boil water before drinking until you hear official word that the water is safe. Report broken water or sewer lines to the proper authorities.

Take extra precautions to prevent fires. Lower water pressure in city and town water mains and the interruption of other services may make fire fighting extremely difficult after a hurricane.

Insurance representatives will be on the scene quickly after a major disaster to speed up the handling of claims. Notify your insurance agent or broker of any losses, and leave word where you can be contacted.

Take steps to protect property. Make temporary repairs to protect property from further damage or looting. Use only reputable contractors (sometimes in the chaotic days following a disaster, unscrupulous operators will prey on the unsuspecting) — check with the Better Business Bureau. Keep all receipts for materials used.

Be patient. Hardship cases will be settled first by insurance representatives. Don't assume that your settlement will be the same as your neighbor's. Policies differ and storm damage is often erratic.

It takes a team effort. Responsibility for cleanup falls to numerous local, state, and federal agencies. A local disaster coordinator will be on hand to help residents in this effort. For more information, contact your county Emergency Management coordinator.

APPENDIX

HURRICANE	YEAR	CATEGORY	DEATHS
Galveston, Texas	1900	4	8,000+
Lake Okeechobee, Florida	1928	4	1,836
Florida Keys/South Texas	1919	4	600
New England	1938	3	600
Florida Keys	1935	5	408
Audrey (Louisiana, Texas)	1957	4	390
North Carolina to New England	1944	3	390
Grand Isle, Louisiana	1909	4	350
New Orleans, Louisiana	1915	4	275
Galveston, Texas	1915	4	275
Camille (Mississippi, Alabama, Virginia)	1969	5	256
Miami, Florida	1926	4	243
Diane (Northeast U.S.)	1955	1	184
Southeast Florida	1906	2	164
Mississippi, Alabama, Florida	1906	3	134
Agnes (Northeast U.S.)	1972	1	122
Hazel (Carolinas, Northeast U.S.)	1954	4	95
Betsy (Florida, Louisiana)	1965	3	75
Carol (Northeast U.S.)	1954	3	60
Floyd (North Carolina, New Jersey)	1999	2	56

Source: NOAA, National Hurricane Center.

THE DEADLIEST
HURRICANES IN THE
UNITED STATES,
1900–1999

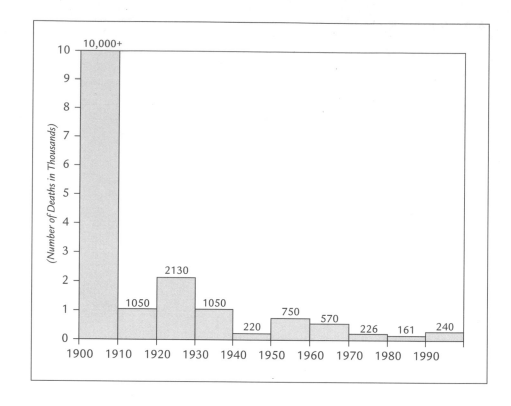

Losses from Hurricanes in the Continental United States, by Decades (through 1999) Source: NOAA, National Hurricane Center.

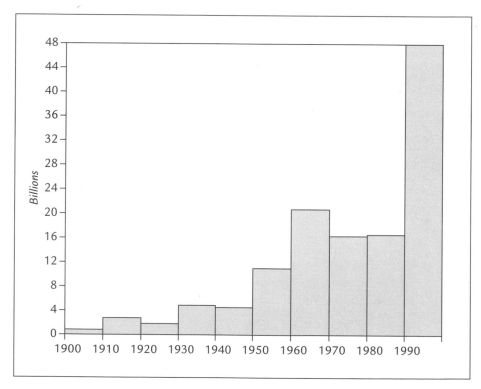

Estimated Losses from Hurricanes in the United States, by Decades (through 1999; damages in 1998 dollars) Source: NOAA, National Hurricane Center.

HURRICANE	YEAR	CATEGORY	DAMAGES ($ BILLIONS)
Andrew (Southeast Florida, Louisiana)	1992	4	26.5000
Hugo (South Carolina, North Carolina)	1989	4	7.0000
Floyd (North Carolina, New Jersey)	1999	2	4.5000
Fran (North Carolina)	1996	3	3.2000
Opal (Northwest Florida, Alabama)	1995	3	3.0000
Georges (Southwest Florida, Mississippi)	1998	2	3.0000
Frederic (Alabama, Mississippi)	1979	3	2.3000
Agnes (Florida, Northeast U.S.)	1972	1	2.1000
Alicia (Texas)	1983	3	2.0000
Bob (North Carolina, Northeast U.S.)	1991	2	1.5000
Juan (Louisiana)	1985	1	1.5000
Camille (Mississippi, Alabama, Virginia)	1969	5	1.4207
Betsy (Florida, Louisiana)	1965	3	1.4205
Elena (Mississippi, Alabama, Florida)	1985	3	1.2500
Gloria (Eastern U.S.)	1985	3	.9000
Diane (North Carolina, Northeast U.S.)	1955	1	.8317
Erin (Florida, Alabama)	1995	2	.7000
Allison (Texas)	1989	T.S.*	.5000
Alberto (NW Florida, Alabama, Georgia)	1994	T.S.*	.5000
Eloise (Northwest Florida)	1975	3	.4900

THE COSTLIEST MAINLAND UNITED STATES HURRICANES, 1900–1999 (DAMAGES IN UNADJUSTED DOLLARS)

*Tropical storm or depression included due to extensive damages
Source: NOAA, National Hurricane Center.

THE COSTLIEST
MAINLAND UNITED
STATES HURRICANES,
1900–1999
(NORMALIZED TO
1999 DOLLARS BY
INFLATION, PERSONAL
PROPERTY INCREASES,
AND COASTAL COUNTY
POPULATION CHANGES)

HURRICANE	YEAR	CATEGORY	DAMAGES ($ BILLIONS)
Southeast Florida, Alabama	1926	4	83.8
Andrew (South Florida, Louisiana)	1992	4	38.4
Galveston, Tex.	1900	4	30.9
Galveston, Tex.	1915	4	26.1
Southwest Florida	1944	3	19.5
New England	1938	3	19.3
Southeast Florida, Lake Okeechobee	1928	4	16.0
Betsy (Florida, Louisiana)	1965	3	14.4
Donna (Florida, Eastern U.S.)	1960	4	14.0
Camille (Mississippi, Louisiana, Virginia)	1969	5	12.7
Agnes (Florida, Eastern U.S.)	1972	1	12.4
Diane (North Carolina, Northeast U.S.)	1955	1	11.9
Hugo (South Carolina, North Carolina)	1989	4	10.9
Carol (Northeast U.S.)	1954	3	10.5
Southeast Florida, Alabama, Louisiana	1947	4	9.6
Carla (Texas)	1961	4	8.2
Hazel (North Carolina)	1954	4	8.2
Northeast U.S.	1944	3	7.5
Southeast Florida	1945	3	7.3
Frederic (Alabama, Mississippi)	1979	3	7.3
Southeast Florida	1949	3	6.8
South Texas	1919	4	6.2
Alicia (Texas)	1983	3	4.7
Floyd (North Carolina, New Jersey)	1999	2	4.5
Celia (South Texas)	1970	3	3.9
Dora (Northeast Florida)	1964	2	3.6
Fran (North Carolina)	1996	3	3.6
Opal (Northwest Florida, Alabama)	1995	3	3.5
Georges (Southwest Florida, Mississippi)	1998	2	3.0
Cleo (Southeast Florida)	1964	2	2.8

Source: "Normalized Hurricane Damages in the United States," by Roger A. Pielke Jr. and Christopher W. Landsea. Updated estimates provided by Chris Landsea of the NOAA Hurricane Research Division.

HURRICANE	YEAR	CATEGORY	PRESSURE (MILLIBARS)	PRESSURE (INCHES)
Florida Keys	1935	5	892	26.35
Camille (Mississippi)	1969	5	909	26.84
Andrew (Florida)	1992	4	922	27.23
Florida Keys, Texas	1919	4	927	27.37
Lake Okeechobee, Florida	1928	4	929	27.43
Donna (Florida)	1960	4	930	27.46
Galveston, Texas	1900	4	931	27.49
Grand Isle, Louisiana	1909	4	931	27.49
New Orleans, Louisiana	1915	4	931	27.49
Carla (Texas)	1961	4	931	27.49
Hugo (South Carolina)	1989	4	934	27.58
Miami, Florida	1926	4	935	27.61
Hazel (North Carolina)	1954	4	938	27.70
Florida, Mississippi, Alabama	1947	4	940	27.76
Texas	1932	4	941	27.79
Gloria (Eastern U.S.)	1985	3	942	27.82
Opal (Florida)	1995	3	942	27.82
Audrey (Louisiana, Texas)	1957	4*	945	27.91
Galveston, Texas	1915	4*	945	27.91
Celia (Texas)	1970	3	945	27.91

THE MOST INTENSE MAINLAND UNITED STATES HURRICANES AT THE TIME OF LANDFALL, 1900–1999

*Classified as category 4 because of estimated winds.
Source: NOAA, National Hurricane Center.

SELECTED NOTORIOUS HURRICANES IN NORTH CAROLINA SINCE 1879

NAME/DATE	CATE-GORY	MAXI-MUM WIND	PRESSURE IN N.C. (INCHES)	N.C. DEATHS	N.C. DAM-AGE (UN-ADJUSTED
August 1879	4	168*	na	40+	na
September 1883	3	100+*	na	53	na
August 1899	4	140*	na	25	na
September 1933	3	125*	28.26	21	$3 million
September 1944	3	110*	27.97	1	$1.5 million
Hazel, 1954	4	150*	27.70	19	$136 million
Ione, 1955	3	107	28.00	7	$88 million
Donna, 1960	3@	120*	28.45	8	$25 million
Diana, 1984	3#	115	28.86	3	$85 million
Gloria, 1985	3	100+*	27.82	1	$8 million
Hugo, 1989	3@	100*	28.88	7	$1 billion
Emily, 1993	3	111*	29.00	0	$13 million
Fran, 1996	3	115*	28.17	24	$5.2 billion
Floyd, 1999	2	110*	28.34	52	$6 billion

*Estimated
@Category 3 in North Carolina; category 4 elsewhere
#Cape Fear area only; was a category 2 at final landfall

NORTH CAROLINA HURRICANES, 1990–1999

NAME	DATE	CATEGORY	MAXIMUM WIND	PRESSURE IN N.C. (MILLIBARS)	N.C. DEATHS	N.C. DAMAGES (UNADJUSTED)
Emily	August 1993	3	111 (g)	982	0	$13 million
Bertha	July 1996	2	108 (g)	973	2	$1.2 billion
Fran	September 1996	3	115	954	24	$5.2 billion
Bonnie	August 1998	2/3	115 (g)	964	1	$480 million
Dennis	August–Sept. 1999	T.S.*	110 (g)	984	0	$100 million
Floyd	September 1999	2	105	981	52	$6 billion

*Tropical storm
Source: NOAA, National Hurricane Center.

Hurricane Evacuation Routes. Source: N.C. Division of Emergency Management.

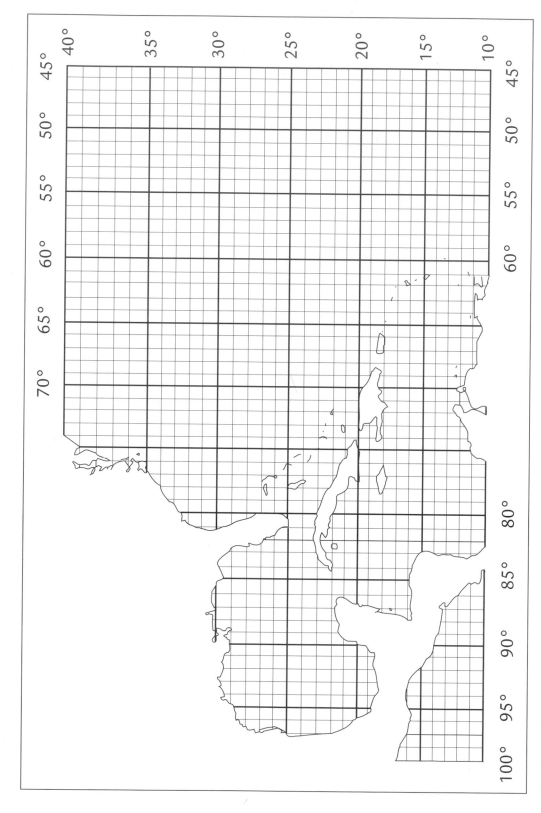

Hurricane Tracking Map

ACKNOWLEDGMENTS

This book was assembled with the outstanding cooperation of many individuals and organizations. Special thanks are offered to those who assisted me in collecting the stories, weather reports, historical data, and photographs. Whenever possible, appropriate credit has been provided for photographic sources.

The primary source of historical and meteorological information used for this text was the National Weather Service Technical Memorandum, *A Historical Account of Tropical Cyclones That Have Impacted North Carolina since 1586*, originally researched by Charles Carney and Albert Hardy and later updated by James Stevenson. This comprehensive publication was used extensively as a resource for details of storm movements, severity, and destruction. Hardy and Carney included numerous uncredited quotations and stories, most of which were borrowed from newspapers and other historical documents. Many of those have been repeated in this book.

Booklets and pamphlets produced by the National Oceanic and Atmospheric Administration (NOAA), the parent organization of the National Weather Service, were the primary resources used in chapters 1, 2, and 3. These excellent publications were useful in providing information on the formation, tracking, and intensity of tropical cyclones.

Chapter 4 relies heavily on the Carney and Hardy publication. Other sources used for this chapter include *Joshua's Dream*, by Susan Carson; a letter from Governor Tryon to Lord Hillsborough from the *Colonial Records, Tryon's Letter Book*; excerpts from the *Raleigh Minerva*; *An Archaeological and Historical Reconnaissance of U.S. Marine Corps Base, Camp Lejeune*, by Thomas Loftfield and Tucker Littleton; and *Graveyard of the Atlantic*, by David Stick.

Weather Bureau records were available for much of chapter 5. Other sources include "An Act of Providence," a widely published article by John Sanders; *The Atlantic Hotel*, by Virginia Doughton; *The Outer Banks* and *Graveyard of the Atlantic*, by David Stick; various articles by Carteret County author Sonny Williamson; reports from the *Raleigh Observer*, the *Beaufort News*, the *Wilmington Messenger*, the *Washington Gazette*, and the *Carteret County News-Times*; and various newspaper reports referenced in articles from the *State* magazine.

Chapter 6 was also compiled with assistance from the Carney and Hardy report, as well as the newspapers mentioned above and the *News and Observer*, the *Greensboro Daily News*, the *Wilmington Morning-Star*, and the *New York Times*; reports from the American Red Cross; letters from the Louis T. Moore Collection at the New Hanover County Public Library; *Sailin' with Grandpa*, by Sonny Williamson; *Ocracokers*, by Alton Ballance; various articles that appeared in *Sea Chest*; "The Great Atlantic Hurricane," an article in the *Hatteras Monitor* by

Rhonda Roughton; "Hurricane Survival on Hatteras," an article in the *State* magazine by Sybil Skakle; several Bill Sharpe articles in the *State*; and *The Hurricane and Its Impact*, by Robert Simpson and Herbert Riehl.

Additional sources used in chapters 7 and 8 include "Hurricane Hazel," an article in the *Tidewater* by Susan Gerdes; *Hurricane Hazel Lashes Coastal Carolinas*, by Wilmington Printing Company, Art Newton, editor; *Making a Difference in North Carolina*, by Ed Rankin and Hugh Morton; various articles from the *State Port Pilot*, the *New Bern Sun-Journal*, the *Coastland Times*, and the *Virginian-Pilot*; an article from the *Duke Power Annual Report of 1989*; the October 1993 issue of the *Hatteras Monitor*; the September 1991 issue of UNC Sea Grant's *Coastwatch*; and personal interviews with Lewis J. Hardee, Tony Seamon, Dorothy Ipock, and other hurricane survivors.

Sources used for the sections on Bertha, Fran, Bonnie, Dennis, and Floyd in Chapter 8 include numerous articles from the *Raleigh News and Observer*, the *Wilmington Star-News*, the *Jacksonville Daily News*, the *New Bern Sun Journal*, the *Goldsboro News-Argus*, the *Outer Banks Sentinel*, the *Miami Herald*, the *Orlando Sentinel*, the *Carteret County News-Times*, the *Wilson Daily News*, the *Virginian-Pilot*, the *Cary News*, and *Weatherwise* magazine; various reports from the National Hurricane Center, the National Weather Service, the National Climatic Data Center, the Hurricane Floyd Redevelopment Center, the North Carolina Division of Emergency Management, the United States Geological Survey, and the *Carolina Skywatcher*; *The Savage Season*, published by the *Wilmington Star-News*; *Bertha and Fran: Coastal Carolina's Stormy Summer of '96*, published by the *Jacksonville Daily News*; *North Carolina Mitigation Strategy Report: Hurricane FRAN*, published by the North Carolina Division of Emergency Management; *Fran*, a collection of essays from LeAnne Smith's seventh-grade class at Topsail Middle School; "Spooky Waters," an essay by Jones Middle School seventh-grader Abby McDonald; "Beach Erosion," an article by Pam Smith in *Coastwatch*; *Hurricane Floyd Relief and Recovery Summary*, a special report from the Hurricane Floyd Redevelopment Center; and various information available via the World Wide Web from the Associated Press, *USA Today*, ABC News, WRAL Online, MSNBC, and *Lowe's Storm 2000*.

Chapter 9 was developed with the support of several National Weather Service publications on northeasters as well as *The Ash Wednesday Storm*, by David Stick. Sources used in chapter 10 include various reports from the newspapers mentioned above; *Reflections of the Outer Banks*, by Donald and Carol McAdoo; and various Weather Bureau records and reports.

Special recognition and thanks are due to several others who assisted with the completion of this publication. They include Max Mayfield, Neil Frank, Bob Sheets, Bob Burpee, Ed Rappaport, Al Hinn, Dan Bartholf, Frank Lepore, and Joel Cline (past and present staff of the National Weather Service); Billy Ray Cameron, Eric Tolbert, Tom Ditt, and Dwayne Moore of the North Carolina Division of Emergency Management; William Gray of Colorado State University; Spencer Rogers of UNC Sea Grant; Bob Williams of the *News and Observer*; Wilmington resident Bob Lane; and Harry Warren of the Cape Fear Museum. Many thanks are offered to those who contributed photographs, especially Jack Goodwin, Roy Hardee, Ed Harper, Hugh Morton, Helen Shore, Scott Taylor, Drew Wilson, Regina Morton, Don Bryan, Randy Davey, John Althouse, Connie Mason, Lisa Taylor, Gary Allen, Keith Greene, Scott Sharpe, Jim Bounds, Robert Willett, and the late Aycock Brown and Art Newton. Others who deserve special thanks are Cathy Piner, Joe Pelissier of the National Weather Service, Stan Goldenberg and Chris Landsea of the NOAA Hurricane Research Division, my mother, wife, and family, and David Perry and the staff of the University of North Carolina Press.

INDEX

Page numbers in italics refer to illustrations.

Adam's Creek, 36
Agnes (hurricane), 18, *135*, 136, 137, 198, 256, 278, 299, 301, 302
Ahoskie, 230
Al's Auto Salvage, 275
Alamance County, 194, 203
Albemarle Sound, 17, 54, 68, 106, 126, 130, 174
Alberto (tropical storm), 278, 301
Aldridge, Eulalia, 257
Allison (hurricane), 1
Alma (hurricane), 162
Altapass, 66
American Humane Association, 274
American Insurance Association, 171
American Insurance Services Group, 212
American Red Cross, 68, 71, 73, 74, 76, 104, 141, *155*, 160, 190, 201, 231, 240, 244, 260
Andrew (hurricane), 10, 21, 31, 32, 156, 157, 199, 222, 263, 284, 301, 302, 303
Anemometer, 10, 14, 176, 177; graph, *14*
Animal Planet, 274
Animals, 37, 50, 52, 70, 186, 187, 191, 192, 197, 208, 232, 237, 246, 249, *250*, 251, 252, 253, 260, *269*, 270–75, 293
Appalachian Mountains, 3, 66
Arapahoe, 72
Artificial reefs, 210, *213*
Ash Wednesday Storm, 8, *10*, *16*, *262*, *264*, 266, 267
Ash Wednesday Storm, The (Stick), 266
Asheville, 42, 99
Associated Press, 171, 211, 233, 240, 279, 282, 285
Atlantic, 219
Atlantic Beach, *17*, 102, *103*, *104*, *106*, 114, 116, *122*, *123*, *125*, *126*, 130, 134, 137, *142*, 148, *161*, 170, 177, 187, 219, 248, *261*, 273, 286

Atlantic Hotel, 42–45
Atlantic Hurricanes (Dunn and Miller), 6
Atlantic Yacht Club, 59
Aurora, 116, 134
Austin, Johnny, 271
Automobile accidents, 107, 117, 130, 132, 142, 172, 199, 203, 219, 234–37, 256–59
Avon, 77, 145, 157, 159
Ayllon, Lucas de, 34

Bald Head Island, 34, 61, 137, 174, 208
Bales, Jerad, 282
Bangladesh, 16
Banks Channel, 58
Barbara (hurricane), 80
Barco, 211
Barnes & Noble, 255
Barrington, Paul, 182
Basnight, Marc, 254
Bass Lake, 194
Bat Cave, 178
Bayboro, 72, 134
Beach nourishment, 248, 285, 286
Beasley, Jean, 187
Beaufort, *11*, 42, 43, 55, 61, 69, 70, 101, *103*, 104–6, 112, 115, 116, 117, 127, 145, *146*, 164, 170, 177, 187, 209, 210, *262*
Beaufort County, 164, 216, 217, 219, 233
Beaufort Inlet, 70
Beaufort News, 70, 71, 270
Beaulaville, 231
Beckwith, George, 273
Belhaven, *19*, 106, *110*, 112, 116, 134, 137, 170, 177, 187, 188, 207, 210, 219
Bell, Debbie, 218
Bermuda, 173
Bertha (hurricane), 1, 3, *7*, 162, *163*, 164, *165*, *166*, 167, 168, *169*, 170, 171, 172, 173, 175, 182, 183, 186, 187, 189,

202, 206, 209, 211, 212, 217, 219, 225, 274, 304
Bertie County, 164, 226
Bethel, 116
Better Business Bureau, 298
Betts, Don, 184
Beulah (hurricane), 20
Bladen County, 131, 141, 188, 228
Blockade Runner Hotel, 225
Blue Cross and Blue Shield of North Carolina, 255
Blue Ridge Mountains, 112, 136, 151, 154
Bob (hurricane), 157, 301
Bogue Sound, *102*
Boiling Spring Lakes, 141
Bonnie (hurricane), 204, *205*, *206*, 207–12, *213*, 217, 219, 225, 246, 304
Boone, 153
Boyer, Sherry, 234, 257
Boyle, Hal, 52
Bradley Creek, 184
Breakers Hotel, 22
Bret (hurricane), 221
Bridge Tender (restaurant), 182
Bridgeton, 219
Broad Creek, 50, 164
Broad River, 136
Brokaw, Tom, 184
Brooks, Bert, 130
Brown, Aycock, 266
Brown, Chris, 257
Brown's Banks, 36
Brunswick Community Hospital, 208, 246
Brunswick County, 17, 47, 82, 88, 95, 106, 137, 141, 142, 153, 156, 167, 168, 175, 179, 180, 202, 206, 208, 216, 224, 226, 228, 231, 246
Brunswick River, 95
Buco, Paul, 257

Burgaw, 115, 189, 197, 258
Burgaw Creek, 236
Burlington, 86
Buxton, 24, 157, 158, 159, 160, 217, *265*
Buyouts, 240, 255, 284, 285

Calabash, 2, 3, 83, 84, 120, 207
Caley, Barbara, 211
Camden County, 207
Camille (hurricane), 2, 3, 10, 20, 21, 23,
    151, 156, 299, 301, 302, 303
Camp Davis, 75
Camp Lejeune, 35, 36, 117, 123, *178*
Canadian Hole, 217
Cantori, Jim, 219
Cape Fear, 164, 181, 206, 207, 214, 225,
    226
Cape Fear Museum, 184
Cape Fear River, 60, 61, 66, 75, 92, 94,
    112, 138, 170, 175, 180, 197, 228, 229,
    230, 236, 245, 247, 251, 273
Cape Hatteras, *12, 21, 40,* 64, 68, 69,
    73, 74, 76, 77, 80, 81, 111, 112, 117, 132,
    134, 137, 143, 144, 145, 157, *160,* 173,
    214, 262, 267
Cape Hatteras Lighthouse, 218, 285
Cape Hatteras School, 159
Cape Henry, Va., 43, 77
Cape Lookout, 10, 43, 46, 65, 108, 116,
    117, 143, 148, 150, 164, 214, 215, 216, 226
Cape Lookout National Seashore, 216,
    217
Cape of Feare, 34
Cape Verde Islands, 114, 123, 147, 149
Carlyle, Terry, 237
Carney, Charles, 34
Carol (hurricane), 80, 81, 82, 108, 193,
    195, 299, 302
Carolina Beach, *iii, ix,* 2, 57, 60, 75, 95,
    96, 97, 98, 111, *139, 140,* 141, 168, 177,
    180, 181, 182, 208, 209, 246, 271, 272,
    273, 285
Carolina Power and Light Co., 137, 181,
    182, 199, 212, 258
Carolina Yacht Club, *59*
Carteret County, 3, 8, 45, 69, 70, 101,
    104, 109, 112, 126, 127, 142, 164, 175,

178, 187, 209, 216, 217, 218, 219, 226,
    247, 248, 273, 286
Carteret County News-Times, 132, 215,
    219
Carter-Finley Stadium, 244
Carthage, 86
Cary, 171, 191, 199, 251
Cashie River, 245
Caskets, 239, 240
Casper's Marina, 164, 187
Castle Hayne, 251
Caswell Beach, 94, 208, 258
Catawba River, 136
Cedar Island, 45, 70, 71, 120, 216, 218,
    270
Chainsaws, 154, 156, 199, 201
Chandler, Badger, 257
Chapel Hill, 175, 190, 192
Charleston, S.C., 47, 151
Charley (hurricane), 23, *146, 147,* 148
Charlotte, 3, 46, 48, 151, *152–54,* 190,
    278, 283
Charlotte News, iii
Charlottesville, Va., 198
Chatham County, 194
Cherry Point, 80, 114, 145, 207, 226
Child, Aaron, 257
Chimney Rock, 179
Chinquapin, 229, 230
Christmas House, 187
Cindy (hurricane), 221
Civilian Conservation Corps, 77
Civils, Elizabeth, 170
Civil War, 38, 42, 169, 184, 239
Clayton, 194, 195, 230, 258
Cline, Joel, 227, 228, 229
Clinton, *12,* 189
Clinton, Bill (president), 171, 202, 253
Clinton, Patti, 191
CNN's *Headline News,* 208
Coastal development, 3, 132, 165, 278,
    281, 285, 286
Coast Guard, 71, 72, 75, 77, 93, 130, 139,
    158, 160, 170, 232, 234, 238, 239, 242
Cole, Ransom, 257
Colonial Beach, 90
Columbus County, 141, 211

Comfort Suites, 170
Congaree River, 136
Congleton, Henry, 44
Connie (hurricane), 4, 19, *106,* 108–10,
    111, 112, 113, 132
Contentnea Creek, 245
Core, Judy, 257
Core Banks, 69, 217
Core Sound, 50, 218
Core Sound sharpie, *51*
Coriolis Effect, 6
Corolla, 3, 218, 271
Crab House (restaurant), 167
Crabtree Creek, *191,* 192, 193, 204
Crabtree Valley Mall, *191,* 192
Craven County, 164, 171, 210, 216, 219,
    231, 233
Crayton, Skip, 275
Cuba hurricane of 1811, 32
Currie, 258
Currituck, 74, 148
Currituck County, 207, 211
Currituck Sound, 130
Cyclones, 6

Dams, 194, 239, 240, 251, 252, 260
Dan River, 136
Danville, Va., 198
Dare, Virginia, 34
Dare County, 147, 148, *155,* 159, 160,
    216, *263*
Davenport, Charles, 238
David (hurricane), 136, 137
Davis, 219
Davis, Brandon, 257
Davis, Donald, 203
Davis, Palmer, 44
Davis Creek, 91
Days Inn, 183, 245
Dead zone, 253
Denning, Kenneth, 257
Dennis (hurricane), 2, 214, *215,* 216–20,
    227, 228, 247, 286, 304
DeYoung, Walt, 204
Diamond City, 45, 50, 271
Diamond Shoals Lightship, 33, 34, 36,
    55

Diana (hurricane), 3, *6*, 18, 23, 137, *138*, *139*, *140*, 141, 142, 144, 146, 172, 209, 304

Diane (hurricane), 4, 19, 107, 108, *110*, *111*, 112, 113, 114, 116, 117, 120, 132, 137, 299, 301, 302

Discussion of Atlantic Basin Seasonal Hurricane Activity in 1995 (Gray et al.), 278

Disosway, Justice, 44

Dixon, Elijah, 72

Dixon Middle School, 167

Do, Thao and Bruce, 190

Donna (hurricane), 3, *11*, *12*, 17, 120, *121–29*, 130, *131*, 132, 137, 172, 271, 272, 302, 303, 304

Donnell, Anne, *140*

Doppler-on-Wheels, 226

Doppler radar, 30

Dorsey, Ricky, 203

Doshoz, S. L., 53, 54

Drake, Sir Francis, 34

Driver, James, 257

Drotar, Ben, 274

Drum Inlet, 69, 116

Duck, 217, 226

Duke Marine Laboratory, 164, 177, 209

Duke Power Co., 155, 156

Dunbar, 233

Dunes Club, *126*, 130

Dunn, 259

Dunn, Gordon, 118

Dunn, John, 44

Duplin County, 141, 179, 189, 197, 203, 210, 231, 232, 237, 251

Durham, 65, 175, 176, 190, 192, 194, 199, 203, 204, 228, 283

Durham County, 175, 194

East Carolina University, 243, 244, 257, 282

Eastgate Shopping Center, 192

Economic Development Administration, 254

Edenton, 106, 131, 210

Edgecombe County, 231, 233, 238, 239, 242, 243, 257

Edna (hurricane), 81, 82, 108

Edouard (hurricane), 173, 174

Efland, 177

Elizabeth City, 77, 106, 116, 126, 131, 188

Elizabethtown, 66

El Niño, 288

Emerald Isle, 102, 148, 162, 170, 177, 187, 209, 210, 226, 247, 248

Emily (hurricane), *3*, *12*, *21*, *31*, 157, *158–61*, 172, 273, *283*, 304

Enfield, 230

Eno River, 192, 194

Erin (hurricane), 1

Extratropical cyclones, 8, 261–67

Faison, 84, 141

Falls Lake, 194, 195, 271

Farmville, 65, 116

Fayetteville, 66, 67, 84, 179

Federal Emergency Management Administration (FEMA), 171, 201, 202, 212, 231, 240, 242, 255, 260; FEMA city, 243

Felix (hurricane), 1, 162

Ferriss, Jason, 193

Fifi (hurricane), 236

Figure Eight Island, 177, 181

First Baptist Church, 183, *185*

Fishing Creek, 230, 259

Flat River, 194

Floods and Droughts: The Worst Are Yet to Be (Heath), 278

Flowers, Cabrina, 238, 257

Flowers, Destiny, 238, 257

Floyd (hurricane), 1, 2, 3, 18, 22, 31, *133*, 216, 220, *221–23*, 224–26, *227*, 228–32, *233–36*, 237–40, *241*, 242–46, *247*, 248–49, *250*, 251, *252*, 253–55, *256*, 257–60, *269*, 274, 275, 278, 279, 282, 283, *284*, 285, 286, 287, 288, 299, 301, 302, 304

Food Lion, 243, 255

Foreman, Patricia, 242

Fort Bragg, 194

Fort Caswell, 64

Fort Fisher, 75, 170, 181, 182, 226, 247

Fort Macon, 43, 44, 71, 109, 130, 210

Foust, Cristina Marie, 203

Fran (hurricane), 1, 2, 3, 10, 11, 17, 22, 31, 32, 156, 162, 172, *173*, 174, 175, *176*, 177, *178*, 179, *180*, 181, 182, *183*, 184, *185*, 186–90, *191*, 192, *193*, 194, *195*, *196*, *197*, 198, 199, *200*, 201–4, 206, 207, 209, 212, 217, 219, 221, 225, 226, 228, 229, 245, 246, 254, 255, 259, 260, 270, 271, 272, 273, 274, 275, *277*, 278, *279*, 280, 283, 284, 286, 287, 301, 302, 304

Fran (Topsail Middle School Students), 172

Frank, Neil, 287

Franklin County, 194

Freeland, 230

Fremont, 251

Frisco, 157, *159*, 207

Frying Pan Lightship, 47

Frying Pan Light Station, 164, 177

Frying Pan Shoals, 207, 216, 226

Fujita Scale, 226

Fuquay-Varina, 194

Galveston hurricane of 1900, 17, 28, 299, 302, 303

Garner, 179

Garrison, Max, 255

Gaston, 259

Gastonia, 153

Georges (hurricane), 206, 302

Gerdes, Susan, 96

German U-boats, 77

Gert (hurricane), 221

Gilbert (hurricane), 22

Ginger (hurricane), 134–36

Gioglio, Len, 186

Gladys (hurricane), 118

Glaxo-Wellcome, 201, 255

Gloria (hurricane), 3, 25, *142*, *143*, 144–46, 172, 217, 301, 303, 304

Godwin, 259

Goldsboro, 37, 65, 84, 86, 131, *133*, 137, 176, 222, 230, 245, 251, *252*, 257, 259

Gomez, Mario, 257

Gomez, Silverio, 257

Gooding, Linwood, 258

Gordon (hurricane), *161*, 217

Grandberry, Randolph, 258
Granville County, 68, 194
Graveyard of the Atlantic (Stick), 55
Gray, William, 204, 227, 288
Great Atlantic hurricane of 1944, 4, 76, 77, 273, 299, 302, 304
Great hurricane of 1780, 205
Green, Leslie, 198
Greene, Georgia, 204
Greene County, 228, 231, 233
Green Island Hunt Club, 73
Greensboro, 204
Greensboro Daily News, 130
Greenville, 18, 116, 131, 164, 177, 207, 212, 228, 229, 230, 231, 232, 235, 243, 244, 251, 258, 259, 278
Greenville All-Stars, 212
Gregory Hotel, 45
Grifton, 245, 251, 258, 259
Grimesland, 257
Guion, Owen, 44
Gulf Majesty, 238, 239
Gulf Stream, 8, 24, 49, 138, 163, 271, 273
Gulf Stream IV, *30*, 286
Gull Shoal Life-Saving Station, 56
Gum Neck, 210

Habitat for Humanity, 255
Hadnot Point, 36
Hagie crop sprayer, 238
Hall, Billy Ray, 254
Halloween Storm of 1991, *16, 264*
Hamo, Evan and Eric, 191
Hamo, Laura, 191
Hampstead, 167, 180
Harbor Island, 182
Hardee, Lewis, 93, 94
Hardy, Albert, 34
Harker's Island, 50, 116, 216
Harlow, 219
Harned, Steve, 279
Harnett County, 175
Harrellsville, 258
Harrison, Benjamin, 258
Harris-Teeter, 255
Hatteras, 43, 45, 51, 52, 77, 157, 159, 216, 217

Hatteras Inlet, 37
Hatteras Inlet Life-Saving Station, 65
Hatteras Island, 74, 170, 207
Hatteras Lows, 262
Haw River, 194
Hayesville, 258
Hazel (hurricane), *iii, vii, ix*, 2, 3, 4, 8, 10, *12, 13, 15*, 17, 18, 22, 23, 48, 57, 80, *82*, 83, *84*, 85, 86, 87, *88, 89*, 90, 91, *92–105*, 106–8, 113, 115, 119, 120, 130, 132, 141, 168, 177, 179, 180, 184, 190, 221, 227, 246, 272, 278, *280*, 286, 299, 302, 303, 304
Healy, Bernadine, 240
Helene (hurricane), 119, 184
Helicopter rescues, 218, 232, *233*, 234, 235, 236, 238, 239, 240, 242, 274
Helms, Connie and Jerry, 90, 91
Hendricks, Lou, 258
Hernandez, Shad, 239
Hertford County, 164
Hewletts Creek, 177
Hickory, 151, 153
High Point, 86
Hilda (hurricane), 118
Hilliardston, 230
Hillsborough, 172, 192, 194, 251
Hobgood, 259
Hobucken, 219, 226
Hodges, Luther H., *113*, 118, 120
Hoffman, Renee, 234
Hofmann Forest, 164
Hog lagoons, 171, 249, 250, 260, 273, 282
Hoke, Gerald, 258
Holden, R. W., 237
Holden Beach, 17, 84, 88, 90, *138*, 156, 181, 208, 246
Holly Ridge, 210
Homestead Air Force Base, 21
Hookerton, 230
Hotel Brunswick, 47
Hot Springs, 258
Howell, Jim, 234
Hubert, 234
Hughes, John, 44
Hugo (hurricane), 2, 3, *5*, 10, 11, 32, 48,

149, 150, 151, 152–55, 156, 157, 172, 175, 188, 190, 209, 212, 224, 273, 274, 278, *291, 295*, 301, 302, 303, 304
Hunt, Jim (governor), 171, 202, 211, *213*, 253, 254, 256, 266, 284
Hunt Horse Complex, 171
Hurricane Alley, 4, 80
Hurricane Floyd Disaster Relief Fund, 255, 260
Hurricane Floyd Redevelopment Center, 253, 254
Hurricane hunters, 29, *30*, 175, 176, 216
Hurricane of: June 1586, 34; September 1667, 35; August 1750, 35; September 1752, 35, 36; September 1761, 36; September 1769, 36; September 1815, 36; June 1825, 36; August 1827, 36; August 1837, 37; October 1837, 37; November 1837, 37; July 1842, 37; August 1842, 37; September 1846, 37; September 1856, 38; November 1861, 38; September 1876, 40–42, 272; August 1879, 42–46, 304; September 1881, 46, 272; September 1882, 46; October 1882, 46; September 1883, 47, 256, 304; August 1885, 47; August 1893, 48; October 1893, 48, 49; August 1899 (*see* San Ciriaco); October 1899, 57, 58, 59, *60*, 61; November 1904, 64; September 1906, 64; September 1913, 64, *65*, 69, 115; July 1916, 66; September 1928, 66; October 1929, 66, *67*; August 1933, 67, *68*; September 1933, 68, 69, *70*, 71–73, 115, 216, 270, 304; September 1936, 73–74; August 1944, *74*, 75; September 1944 (*see* Great Atlantic hurricane of 1944)
Hurricane prayer, 114
Hurricane Rock: wreck of, *183*, 184
Hurricanes: formation of, 6–8; cross-section of, 7; eye, 7, 141, 145, 157; season, 8; frequency, 8, 40; winds, 10–14, 24; timber losses, 11, 153, 154; storm surge, 14–17; rainfall, 18; evacuations, 23, 110, 111, 142, 143, 144, 148, 157, 174, 223, 224, 305; forward speed,

24, 25; forecasting, 28, 279, 286, 287; warnings, 28–30; naming of, 31, 32; preparations for, 117, 291–98; longest lived, 134; recovery, 284–88; deadliest, 299; losses, 300; costliest, 301, 302; most intense, 303; notorious in North Carolina, 304; tracking map, 306
Hurricane warning, 188, 189
Hurricane watch, 187, 188
Hyde, Frank, 71
Hyde County, 41, 126, 148, 173, 219, 272

IGA supermarket, 245
Indian Beach, 248
Indian Beach Pier, 209
Inlet formation, 3, 17, 37, 69, 186, 247
Intracoastal Waterway, 183
Ione (hurricane), 4, 19, *113*, 114, *115*, 116, 117, *118*, *119*, 132, 304
Iredell County, 136
Irene (hurricane), 229
Iris (hurricane), 1
Iron Steamer Pier, 209

J. H. Rose High School, 235
Jackson, Jesse, 240
Jacksonville, 35, 120, *129*, 164, 167, 179, 188, *206*, 207, 226, 251
Jacksonville Daily News, 7, 166, 167, 169, 173, 176, 178
Jake's Barber Shop, 245
Janet (hurricane), 118
Jarrell, Jerry, 204
Jarvis, Thomas J., 42, 44, 45
Jefferson, George, 258
Jenkins, Ossie Lee, 258
Jet-Ski, 188, 194, 197, 237, 249
Jet stream, 29
John F. Kennedy, 239
Johnny Mercer's Pier, 120
Johnson, Tim, 187
Johnston, 35
Johnston County, 175, 195, 258
Jones, David, 166, 167
Jones, Ernita, 258
Jones, George, 258

Jones, Mattie, 243
Jones, Osseynna, 258
Jones, Reginald, 258
Jones, Tom, *165*
Jones, William, 258
Jones County, 164, 211, 231, 232, 245, 275
Jones Middle School, 220
Jordan Lake, 271, 272

Keith, Robert, 234
Kellam, Violet, 131
Kelly, 189
Kemp, Rodney, 271
Kenansville, 189, 231
Kenly, 251
Kennedy Space Center, 223
Kernersville, 48
Key West, Fla., 1
Kill Devil Hills, 266
King's Barbeque (restaurant), 245
Kinston, 84, 115, *131*, 164, 195, 196, 197, 228, 229, 230, 231, 232, 245, 251, 255, 258
Kitty Hawk, *16*, 43, 45, 131, 217, 218
Kitty Hawk Elementary School, 218
K-Mart, 242
Kure Beach, 164, 168, 181, 182, 207, 226

Labor Day storm of 1935, 21, 22, 123, 303
Lake Lure, 178
Lake Raleigh, 194
Lane, Bob, 169
Lane, Ralph, 34
La Niña, 204, 288
Latham, Derek, 235, 236
Leland, 257
Lenny (hurricane), 221
Lenoir County, 231
Lexington, 86
Life-Saving Stations, *49*, 54, 56, 57, 64, 65
Lincolnton, 153
Little Bridge, 131
Little League World Series, 212
Little Rascals, 275
Little River, 84, 87

Little River Creek, 258
Little Swift Creek, 245
Logan, Cecil, 181
Lola, 70
Long Beach, 17, *18*, *82*, 88, *89*, 90, 91, 119, *221*, 156, 181, 246, *247*
Long Beach Pier, 246
Looting, 99, 100, 240
Lost Colony, The 128
Louisburg, 230
Lowe's Home Improvement Warehouse, 255
Lucama, 230
Luis (hurricane), 1
Lukens, 70
Lumber River, 136
Lumberton, 66
Lyons, Steve, 215

McArthur, Faye and Paul, 272, 273
McDonald, Abby, 220
McEachern, Alex, 96
Magnolia, 236
Making a Difference in North Carolina (Rankin and Morton), iii
Maldonado, Eusebio, 258
Mandarino, Antonio, 211
Maness, Ed, 202
Manteo, 73, *128*, 145, *147*, 148, 210, 266
Maple Hill, 198
March Superstorm of 1993. *See* Storm of the Century
Marilyn (hurricane), 1
Marine fisheries, *109*, 253, 272, 273, 274
Marshallburg, 50
Marsh hen tide, 84
Martin County, 164
Masonboro Island, 226
Mason Inlet, 182, 246
Mattamuskeet, 115
Maxton, 258
Mayfield, Max, 287, 288
Mayo, Ben, 238, 258
Mayo, Keisha, 238, 258
Mayo, Vivian, 238, 258
Maysville, 114, 210, 251
Mecklenburg County, 153, 156

Mencias, Erasmo, 236
Merrimon, 70, 71
Miami Herald, 224
Middle Creek, 194
Midgett, Rasmus, 56
Mills, David, 258
Mitch (hurricane), 205, 206
Mitchell, Jake, 245
Mobley, Emily and Paul, 258
Monroe, 153
Moody, Marvin, 258
Moore, Robyn, 208
Morehead City, *12*, 44, 45, 50, 61, *70*,
    101, *102*, 103–5, 114, 115, 116, 120, *124*,
    127, *143*, 144, 145, 170, 210, 266, 272
Morehead City Hall, *124*
Mount Mitchell, 18, 263
Mount Olive, 257, 259
Mount Olive Pickle Company, 255
Mount St. Helens, 151
MSNBC News, 237
Myers, Joe, 224
Myrtle Beach, S.C., 48, 64, *79*, *81*, 87,
    151, 174, 226, 228

Nags Head, 7, 54, *68*, 73, 77, 80, 131, 144,
    158, 172, *215*, 217, 218, *264*, *265*, 272
Nash County, 141, *233*, 234, 258, 259
Nashville, 259
National Climatic Data Center, 171,
    203
National Flood Insurance claims, 260
National Guard, *89*, *97*, 98, 100, *110*,
    160, 196, 197, 202, 208, 210, 217, 218,
    232, 242, 244, 245, 252, 258
National Gypsum Company, 255
National Hurricane Center, 1, 21, 30,
    31, 157, 162, 164, 174, 177, 204, 207, 212,
    214, 219, 220, 221, 254, 256, 257, 278,
    286, 287, 288
National Severe Storms Laboratory,
    176
National Weather Service, 4, 84–86,
    112, 114, 134, 141, 144, 145, 148, 170, 171,
    175, 177, 179, 190, 216, 226, 227, 228,
    230, 245, 263, 279, 292
Neuse River, 52, 64, 68, 69, 170, 171,

192, 194, 195, 196, 197, 216, 226, 228,
    229, 230, 236, 244, 245, 251, 258, 273
Neuse Sport Shop, 245
New Bern, *19*, 36, 45, 61, 64, 65, 69, 106,
    112, *115*, *118*, *119*, 131, 145, 170, 177, 187,
    225, 255, 275
New England hurricane of 1938, 76
New Hanover County, 1, 95, 112, 137,
    141, 156, 167, 175, 181, 184, 201, 206,
    208, 216, 225, 258
New Hanover Regional Medical Cen-
    ter, 208
Newport, 112, 115, 226, 272
Newport River, 272
New River, 35, 36, 61, 137, 188
New River Air Station, 226
News and Observer, 116, 117, 145, 165,
    170, 176, 179, 180, 183, 184, 185, 187,
    189, 190, 191, 192, 193, 195, 196, 197,
    200, 202, 212, 218, 219, 227, 234, 236,
    240, 242, 243, 257, 272, 274, 282, 284
New York Times, 69
Nichols, Reiford, 258
Nixon, William, 258
Norfolk, Va., 1, 55, 68, 225
Norlina, 258
Norman, Leatha, 233
North Carolina Aquarium, 182
North Carolina Department of Agri-
    culture, 189
North Carolina Department of Trans-
    portation, 227, 231, 259
North Carolina Division of Emer-
    gency Management, 194, 203, 234,
    240, 249, 254, 287, 292
North Carolina Division of Forest
    Resources, 201
North Carolina Division of Marine
    Fisheries, 210, 274
North Carolina Division of Water
    Resources, 283
North Carolina General Assembly,
    254
*North Carolina Herpetological Society
    Newsletter*, 187
North Carolina Highway Patrol, 99,
    232, 237

North Carolina Museum of Art, 275
North Carolina Press Association, 42
North Carolina Sea Grant, 286
North Carolina state medical exam-
    iner, 256, 257, 260
North Carolina State Port, *143*, 145
North Carolina State University, 172,
    190, 270, 275
North Carolina Wildlife, 273
Northeasters, 7, 8, *9*, *10*, 261–66, 267,
    282
North River, 106, 112, 115, 219
North Topsail Beach, 164, 165, *166*,
    167, *169*, 173, 184, 186, 202, 203, 209,
    223, 247
Norton, Grady, 83

Oak Island, 75, 84, 139, 221, 226, 246,
    247
Ocean Crest Pier, 246
Ocean Isle, 17, *88*, 90, 91, 92, 156, 181
Ocracoke, 24, 29, 36, 41, 45, 51, 61, 67,
    72, 73, 76, 117, 144, 145, 148, 157, 170,
    216, 218, *265*, 272, 273, *291*, *295*
Ocracokers (Ballance), 273
Old Stump Inlet, 61
Old Town Point, 35
O'Neal, Bud, 219
Onslow County, 1, 35, 41, 61, 156, 164,
    167, 175, 178, 179, 203, *206*, *213*, 225,
    226, 247
Opal (hurricane), 1, 3, 203, 301, 302, 303
Orange County, 258
Oregon Inlet, 37, 150, 216
Oriental, 72, 170, 210, 219, 273
Otway, 116
Outer Banks, 3, 4, 16, 17, 18, 21, 24, 25,
    29, 30, 34, 37, 49, 54, 67, 81, 116, 125,
    130, 143, 146, 148, 157, 161, 162, 207,
    214, 217, 218, 220, 224, *264*, *265*, 266,
    273, *282*, *283*

Pactolus, 238, *241*
Pamlico County, 126, 216, 219, 226
Pamlico Queen, 170
Pamlico River, 116, 170, 188, 216, 219,
    226

Pamlico Sound, 17, 68, 106, 112, 130, 134, 137, 148, 158, 159, 174, 218, 226, 253, 273
Parkton, 106
Pasquotank County, 207
Pasquotank River, 188
Pee Dee River, 136
Pembroke, 258
Pender County, 1, 141, 156, 164, 168, 175, 179, 181, 197, 201, 204, 225, 231, 236, 237, 257
Penderlea, 109
Penland, Leon, 258
Perfect Storm, The (Junger), 239
Perquimans County, 226
Person County, 175
Phelps, Freddy, 209
Phillips, Richard, 259
Piers, *21*, 116, *142*, 156, 164, 167, 209, 246, 273
Pine beetles, 187
Pine Knoll Shores, 127, 187, 209, *213*, 248, *277*
Piner, Mitchell, 236, 259
Pinetops, 238, 257, 258, 259
Pink Hill, 210, 251
Pitt County, 164, 231, 232, 233, 238, 244
Pittman, Lupton, 192
Pitt Memorial Hospital, 244
Plymouth, 116
Pollocksville, 231, 245
Portsmouth, 29, 36, 43, 45, 50, 61, 65, 271
Powell, Mark, 175
Poythress, Charlotte, 259
Princeville, 232, 237, 239, 240, *241*, 243, 255
Priscilla: wreck of, *39*, 55, *56*, 57
Promised Land, 50
Providence Methodist Church, *41*
Pungo River, 170, 188, 210, 219

Qian, Song, *195*
Quackenbush, Tom, 193, 194
Queen, Mike, 183, 184

R. J. Reynolds Tobacco Co., 255
Racer's Storm, 32, 37
Railroads, 14, 44, 46, 64, 65, 69, 130
Rainfall, record, 66–67, 114, 228
Raleigh, Sir Walter, 34
Raleigh, 3, 48, 64, 84, 86, 89, *105*, 137, 171, 175, 176, 179, 188, 190, 191, 192, *193*, 194, *195*, *196*, *197*, 199, *200*, 202, 203, 204, 228, 243, 244, 271, 272, 275, 278, 279
Raleigh-Durham Airport, 84, 112, 177, 179
Raleigh Minerva, 36
Raleigh Observer, 42
Raritan River, 230, 248
Reaves, Mary Bland, 192, 203
Red Crescent, 118
Redwine, David, 285
Reece, Roger, 193
Reentry passes, 142
Reflections of the Outer Banks (McAdoo and McAdoo), 271
Reid, Otis, 259
Research Triangle Park, 1
Reuters News Service, 198
Rhodes Livestock Farm, 171
Richlands, 167, 219, *235*
Riegelwood, 251
Riggs, Stan, 282
Right-front quadrant, 23, *24*, 25
Roanoke colony, 34
Roanoke Island, 34, 37, 134, 207
Roanoke Sound, 131
Robbins, 86
Robeson County, 258
Robinson's Beach, 88, 90
Rockfish Creek, 236, 237
Rock River, 136
Rocky Mount, 18, 41, 194, 228, 229, 230, 231, 232, 237, 243, 255, 257
Rodanthe, 217, 218
Roe, 70, 270, 271
Rogers, Josh, *165*
Rogers, Spencer, 286
Rogers Bay Campground, 209
Rolling View Marina, 194
Rose, George, 168

Roseboro, 141
Rose Hill, 203
Rouse, Marion, 203
Russell, Ronald, 259
Russian Red Cross, 118

Sadler, Sam, 41
Saffir-Simpson scale, 20, 21, 22, *23*
St. James Plantation, 208
Salter Path, 50, 102, 112, 114
Saluda River, 136
Salvation Army, 160, 201
Sampson County, 131, 141, 175, 189, 211
San Ciriaco hurricane of 1899, 4, 49, 50–54, *55*, *56*, 57, 61, 270, 273, 304
Sandy Point, 219
Sanford, 251
Sanitary Restaurant, 104
Santa Ana storm of 1825, 32
Satellite images, *5*, *6*, *161*, *135*, *138*, 221, *284*
Savage Season, The (Wilmington Star-News), 166
Sawyer, Chris, 238
Schafer, Dennis, 275
Scheer, Julian, *iii*
Schindelar, Henry, 187
Scotland Neck, 259
Scotts Hill Marina, 184, 204
Seacoast train, *58*
Sea Islands hurricane of 1893, 48, 272
Sea Level, 116, 219
Seamon, Tony, 103, 104
Sears, Steve, 203
Seven Springs, 245
Shackleford Banks, 45, 50
Shallotte, 207, 246
Shallotte River, 246
Sharpie, *51*
Shell Island Resort, 182, 246
Shenandoah Valley, 198, 220
Sheraton Crabtree Hotel, *191*, 192
Shipwrecks, 37, 46, 47, 48, 52, 54, 55, *56*, 57, 61, 64, 65, 77, 92–95, *105*, 109
Showboat Motel, 161
Silver Lake Waterpark, 194

Smallwood, Ather, 259
Smartley, Freddie, *205*
Smith, Lonnie, *234*
Smith, Roger, 259
Smith, Wayne, *193*
Smithfield, 195, 251
Smithville. *See* Southport
Smyrna, 45
Snakes, 208, 210, 237, 246, 272, 275
Snap Dragon (restaurant), 167
Snead's Ferry, 100, *101*, 115, *165*, 252
Snow, 64, 262, 263
Snow Hill, 164, 228
Snow's Cut, 168, 184
Southern Baptist Convention, 160
Southport, 36, 40, 41, 46, 47, 57, 61,
    74, 75, *88*, 91, *92–95*, 119, *140*, 141, 168,
    177, 181, *205*, 208, 216, 266
South River, 70, 72, 219
Soward, Granger, 273
Spencer, Larry and Ginny, 168
"Spooky Waters" (McDonald), 220
Sportsman's Pier, 273
Spring Hope, 257, 258
Springsteen, Benjamin E., 55–57
Stacy, 219
Starzynski, John, *223*
State Port Pilot, 90
Stevens, Corrie, *215*
Stevenson, James, 34
Stick, David, 55, 266
Stockroom Shoes, 255
Stokes, James, 259
Stone's Bay, 36
Stony Creek, 259
Storm of the Century (March 1993), 8,
    261, *262*, 263, 266
Storm surge, 14, *15*, 16, 17
Strickland, Alice, *97*
Stumpy Point, 210
Suarez, Candido, 188
Subsidence, 283
Sugar Loaf Mountain, 179
Summerlin, Larry, 259
Sunny Point Military Ocean Termi-
    nal, 94, 247
Sunset Beach, 17, 208

Super 8 Motel, 245
Supply, 208, 246
Surf City, 17, 164, 166, 167, 184, 185, 204,
    209
Surry County, 136
Swan Island, 50–52
Swan Quarter, *41*, 131, 144, 207, 210
Swansboro, 100, 115, *125*, *163*, 164, 167,
    177, 187
Sweeting, Don, 211

Tarboro, 18, 46, 65, 228, 230, 232, 237,
    240, 242, 243, 255
Tarboro High School, 241, 242
Tar River, 194, 228, 229, 230, 236, 239,
    240, *241*, 243, 245, 251
Taylor's Creek, 105, *146*
Tick Creek, 194
Tidewater, 96
Tolbert, Eric, 225, 226, 227, 254, 285,
    287
Topsail Beach. *See* North Topsail
    Beach; Topsail Island
Topsail Island, 100, *101*, 125, 134, 163,
    164, 165, *166*, 167, 168, *169*, *173*, *176*,
    177, 184, 185, 186, 187, 201, 202, 204,
    209, 214, 226, 247, 271, 274, *279*
Tornadoes, 6, 19, 20, 131, 136, 137, 141,
    171, 179, 180, 210, 220, 226, 263, 288;
    record activity, 20, 162
Torrez, Larry, *222*
Town Creek, 238
Trenton, 220, 230, 245, *250*, 259
Trent River, 69, 245
Triangle. *See* Chapel Hill; Durham;
    Raleigh
Triple S Pier, 102
Tropical depression, 6
Tropical storm, 7
Tropical wave, 6
Typhoons, 6
Tyrrell County, 126, 210

U.S. Air Force, 172, 176, 177
U.S. Army, 173, 235
U.S. Army Corps of Engineers, 195,
    217, 226, 240, 267, 285, 286

U.S. Geological Survey, 86, 194, 195,
    217, 229, 230, 279, 282
U.S. Marine Corps, 100, *123*, *233*, *234*,
    235
U.S. Navy, 170, 232, 235, 236, 238,
    239
U.S. Senate, 118
U.S. Small Business Administration,
    260
U.S. Tobacco Co., 201
University of Miami, 244
University of Oklahoma, 226

Vanceboro, 245, 257
Vandemere, 72
Vause, Ann, *169*
Verrazano, Giovanni da, 34
Vines, Teshika, 238, 259
Virginia Beach, Va., 198, 211
Virginian-Pilot, 131, 215, 231, 234, 245

Wake County, 175, 192, 194
Wake Forest, 251
Wallace, 106, 189, 236, 237, 257, 259
Walnut Creek, *252*
Walt Disney World, 224
Wantland's Ferry, 35
Warren, Bob, *169*
Warren, Curtis Wayne, 203
Warren, Harry, 184
Warren County, 86, 258
Warrenton, 258
Warsaw, 106, 189
Washington, D.C., 198, 202
Washington, N.C., 36, 64, *65*, 106, *108*,
    112, 115, 116, 131, 164, 170, 177, 188, 207,
    210, 219, 225, 231
Washington Gazette, 52
Waterfowl, 41, 271
Water Resources Research Institute,
    254; *News*, 283
Waterside Theater, *128*
Waterspouts, *20*
Wathen, Elaine, 240
Watts, Rolonda, 167
Wayne County, 195, 251
Waynesboro. *See* Goldsboro

Weather Bureau, 28, 40, 89, 117, 118, 120, 132; creation of, 28
Weather Channel, 208, 215, 219
Webb, Thomas, 45
Wenona, 210
Westry, Artemus, 259
White, John, 34
White Oak, 230
White Oak River, 167
Whiteville, 91, 258
Whitley, Cheryl, 259
Whittington, Sarah, *94*
Whittler's Bench, 168, *205*
Wilde, Matt, 237
Wilder, James, 259
Williams, Gary, 236
Williams, Ray, 192
Williamson, Sonny, 71
Williamston, 145
Wilmington, 18, 37, 41, 46, 48, 57, 60, 75, 80, 84, 94, 95, 109, 111, 119, 125, 141, 148, 162, 163, 169, 177, 183, *185*, 188, 192, 202, 206, 207, 208, 226, 228, 237, 247, 257, 271
Wilmington Beach, *22*, *98*
Wilmington Hilton, 170
Wilmington Messenger, 58, 61
Wilmington Morning Star, 75
Wilmington Star-News, 166, 168, 181, 182, 184, 185, 186, 198, 209
Wilson, James, Jr., 259
Wilson, Nat, 283
Wilson, 65, *105*, 251, 257
Windjammer Restaurant, 246
Winds. *See* Hurricanes: winds
Windsor, 245, 258, 259
Winston-Salem, 36
Witt, James Lee, 202
Woods, James, 192
Woodson, 219

World Meteorological Organization, 32
World Wide Web, 174
WRAL News, 254, 275
Wreck Point, 43
Wright Memorial Bridge, *143*
Wrightsville Beach, *vii*, 17, 38, 46, 57, 58, *59*, *60*, 64, *67*, 75, 80, *84*, 98, *99*, *100*, 120, 137, 139, 145, 162, 163, 164, 168, 177, 181, 182, *183*, 199, 207, 208, 209, 214, 216, 225, 246, 266, 280, 285
Wrightsville Beach Yacht Club, *vii*

Yadkin River, 18, 136
Yarborough House Hotel, 45
Yates Millpond, 194
Yaupon Beach, 246

Zebulon, 204, 228, 251, 258
Zeke's Island, 61